WHY

SIZE

MATTERS

WHY

SIZE

MATTERS

FROM BACTERIA TO BLUE WHALES

John Tyler Bonner

PRINCETON UNIVERSITY PRESS
Princeton & Oxford

Published by Princeton University Press, 41 William Street,
Princeton, New Jersey 08540
In the United Kingdom: Princeton University Press,
99 Banbury Road, Oxford OX2 6JX
press.princeton.edu

All Rights Reserved

First paperback printing, 2012
New paperback printing, 2024
Paper ISBN 978-0-691-25440-1
ISBN (e-book) 978-0-691-25442-5

The Library of Congress has cataloged the cloth edition of this book as follows
Bonner, John Tyler
Why size matters : from bacteria to blue whales / John Tyler Bonner.
p. cm.
Includes bibliographical references and index.
ISBN-13: 978-0-691-12850-4 (hardcover)
ISBN-10: 0-691-12850-2 (hardcover)
1. Body size. I. Title.
QL799.B66 2006
578.4'1—dc22   2006004945

Cover design by Michael Boland for TheBolandDesignCo.com
Cover image: Wirestock / iStock

British Library Cataloging-in-Publication Data is available
This book has been composed in Bembo and Helvetica Neue

for Slawa

CONTENTS

CONTENTS

One can live in the shadow of an idea
without grasping it.

—*Elizabeth Bowen*

PREFACE

Our interest in the size of things is entrenched in the human psyche. It reveals itself in literature from *Gulliver's Travels*, to the Grimms' fairy tales, to *Alice in Wonderland*. We see it in our daily thoughts of our growing children, of the people who are around us, of our pets, of the fish we catch, of the portions of the food we are served, of the clothes we buy—are you small, medium, or large?—and one could go on and on. There is hardly anything we observe in daily life that we, either consciously or unconsciously, do not take measure of its size. We love to measure everything with rulers and scales and clocks.

I began to think of the matter of biological size years ago when I first read that glorious chapter in D'Arcy Wentworth Thompson's *On Growth and Form* called "On Magnitude." It is a model of insight, erudition, and beautiful prose. He showed me that size and shape are indeed interrelated and that the reason that this is so is a matter of physics that underlies the biology. From this initial inspiration there slowly grew inside me the feeling that there was a hidden other dimension of the subject that was eluding me. That inner feeling persisted for many years, and slowly something began to take shape. I am finally putting it all together in this book—I feel as though those shadowy thoughts have erupted through the surface.

This book is a summary of those thoughts. It is an enormous subject that I try to bring down to reasonable dimensions so that I can include it all. As will be clear, I am interested in painting the big picture on a small canvas.

If we are a bacterium, or an elephant, or a human being, we have our own size worlds, and for each of us there are things smaller and larger than ourselves. But no one can escape the universal rules imposed by size.

In looking at the subject of biological size in its entirety, from large to small, from plant to animal to microbe, it will be evident that everything is interconnected. An examination of the effects of size is a way of bringing all life together.

Just as the content of this book has almost taken a lifetime to mature, the actual writing has been an equally painful and slow process with innumerable adjustments and corrections in my course as I proceeded. These were greatly helped by the kindness and wisdom of numerous individuals to whom I am deeply indebted. Before I even began the book, my colleague Henry Horn was enormously helpful (as he has always been over the years) in purifying my thoughts about size. At a very early stage of writing I had the help of my friend Jonathan Weiner, who urged on me the need for a sense of direction. The first complete draft of this book (still in its underwear!) was prematurely sent to two anonymous readers, and while the comments of one of them were highly critical, they gave me the needed jolt at just the right moment. Later drafts were greatly improved due to the comments of Sam Elworthy, Brian Hall, Slawa Lamont, Mary Jane West-Eberhard, and another anonymous reviewer. Also I want to thank David Kirk for his help with the section on *Volvox* and its relatives, and my colleague Ted Cox on some matters of physics. I almost feel as though they all ought to be listed as co-authors. My special thanks to Alice Calaprice and Deborah Tegarden for their skill and great help in seeing the book through its final stages of preparation. Finally I would like to thank Hannah Bonner for applying her superb illustrator skills to produce seven of the original drawings.

MARGAREE HARBOUR
Cape Breton, Nova Scotia

WHY

SIZE

MATTERS

CHAPTER I

INTRODUCTION

In the seventeenth century it was held by some that inside a human sperm there was a minute human being—a homunculus—that was planted inside the womb. Development consisted of the miniature homunculus enlarging and passing through birth and on to maturity—just like inflating a balloon. There were others, going back to the early ideas of Aristotle and the many who followed him, who took the view that vast changes in shape occurred between egg and adult, for it could be plainly seen that the early stages of development of any animal bore no resemblance to what came later. These two views frame the point I want to make in this book. In the case of the homunculus, shape is totally unconnected to size; as size increases shape remains unaltered. In the other case—now totally accepted—as size increases from egg to adult, the shape must change; there is no alternative.

Let me put the matter in another way. If an engineer is commissioned to build two bridges, one across the Hudson River and the other across a brook no more than 30 feet wide, it is quite obvious that the two bridges will be very different in their appearance. Even more importantly, they will differ in their construction and materials. These differences will have nothing to do with the artistic whims of the engineer, at least for the larger bridge: they are absolute requirements. Any attempt to build the Hudson River bridge with wooden planks would collapse into the water long before it was finished. The elaborate steel trusses and the carefully designed architecture of the huge bridge are demanded by the width of the Hudson—it is dictated by its large size. As we shall see, this perfectly mirrors what happens in living organisms; they too cannot escape the conditions set by size; they have no choice.

With these thoughts in mind, let me state the main argument of this book. Changes in size are not a consequence of changes in shape, but the reverse: changes in size often require changes in shape. To put it another way, size is a supreme regulator of all matters biological. No living entity can evolve or develop without taking size into consideration. Much more than that, size is a prime mover in evolution. There is abundant evidence for the natural selection of size, for both increases and decreases. Those size changes have the remarkable effect that they guide and encourage novelties in the structure of all organisms. Size is not just a by-product of evolution, but a major player. Size increase requires changes in structure, in function, and, as we will see, in other familiar evolutionary innovations. It requires them because they are needed for the individual to exist. Life would be impossible without the appropriate size-related modifications.

The subject of size has not been ignored in the past. Quite to the contrary, and as will be clear in the pages to come, there is a great literature on matters of size, beginning with the Greeks and bursting into flower with Galileo. This is true for the West, and no doubt there are similar traditions in other cultures.

However, the subject is always to some degree fragmented because it is generally introduced as an adjunct to some other biological phenomenon or property. For instance, the topic might be running speed, or rate of metabolism, or one of many other possibilities, and in the discussion of each of these phenomena the crucial role of size would be included. Many of the themes treated in this book can be found elsewhere. Here I wish to look at them from a different point of view—from the other end of the telescope—and show that the biological world revolves around size.

The mindset that size is not a central issue is quite understandable. To say an elephant is big says nothing about all the things that make an elephant: its anatomy, its physiology, and even its behavior. These are the aspects that draw our attention and the matters we want to study. Yet size is an overarching issue. Its effect is something that no organism, from the smallest bacteria to the largest whale, can escape. It governs their shape and all their activities in a way that is of fundamental significance. Size dictates the characteristics of all living forms. It is the supreme and universal determinant of what any organism can be and can do. Therefore, why is it a subject that always resides in the wings rather than center stage?

The main reason is that organisms are material objects while size is a bloodless geometric construct. Any object, whether animate or inanimate, will have a size. Airplanes, boats, or musical string instruments vary in size just like animals and plants, and in all cases their size and their material construction are totally different matters even though they affect one another.

That the role of size has been to some degree neglected in biology may lie in its simplicity. Size may be a property that affects all of life, but it seems pallid compared to the matter which makes up life. Yet size is an aspect of the living that plays a remarkable, overreaching role that affects life's matter in all its aspects. It is a universal frame from which nothing escapes.

There are many things one wants to know about size, in particular those that concern its evolution. For instance, what is the evidence for my contention that size differences are a prime object of natural selection and are followed by changes in construction? What is the relation between size and internal complexity—that is, the division of labor—and, again, what is the evidence for which came first? What is the relation between size and the timing of all living activities such as the speed of movement of animals, or life span; and does size impose the timing, or the reverse? As we shall see, it is generally true that size is the prime mover: if size changes occur through the agency of natural selection, all those other matters must follow.

SIZE RULES

In the pages to come we will see many examples of where size rules life. They are supported by correlations in which various properties of organisms vary with size. It is these correlations that provide the foundation, the underpinning, for my contention that size rules life. The correlations can be stated in the form of five rules that will be briefly mentioned here and expanded and explained later. The rules are as follows:

RULE 1 Strength varies with size.

RULE 2 Surfaces that permit diffusion of oxygen, of food, and of heat in and out of the body, vary with size.

RULE 3 The division of labor (complexity) varies with size.

RULE 4 The rate of various living processes varies with size, such as metabolism, generation time, longevity, and the speed of locomotion.

RULE 5 The abundance of organisms in nature varies with their size.

Each of these rules will be put in its proper context. Some are physical or engineering principles; rules that apply for size differences occur in the inanimate as well as the animate. A central issue is the role of size in evolution, and this can be seen in numerous manifestations. Size also affects in many fundamental ways the physiology of animals, plants, and other organisms; in fact, this is true for all aspects of living things that involve time or rates of activity. And the human interest in the matter of size (including my own) will not be neglected.

CHAPTER 2

THE HUMAN VIEW OF SIZE

It is only natural that we should measure everything in the world around us in terms of our own size. An elephant is bigger than we are, and a mouse is smaller. Some years ago, Time-Life was publishing a series of illustrated books on various subjects, and they called me up to ask for advice on a book they were doing on the general subject of growth—would I please come in to New York to discuss the matter with their editors, for they had numerous questions they wanted to ask an interested biologist. I no longer remember what those questions were, but at the end of our conference they said they were having difficulty thinking of a suitable photograph for the cover of the book. I thought about it as they were talking, and suggested they should have the large open hand of a man and in his palm have the hand of an infant. They did not seem very enthusiastic about the idea, but when they sent me the finished book, that was exactly what they had on the cover. We see and are conscious of the size of everything that surrounds us, whether it is smaller or larger, and nothing makes the point more clearly than the growth of children. Who has not remarked upon seeing—after an absence—the child of a friend or family, "My, you are so much taller than when I saw you last." Once I visited Louis Pasteur's house outside of Paris, and one of the doorways still had pencil marks recording the annual growth of his children, something that will sound familiar to everyone. Either consciously or, because it is so much part of our natures, unconsciously, we are forever taking note of the size of things and gauging any increase or decrease.

Our world is the world we see with our naked eyes, and that is what we use for our everyday measuring stick. We are also aware that there are worlds that are larger and smaller than we can see in our normal existence—in fact, how to see the things above and below our vision was among the great discoveries in our history. The telescope was one of the profound technical advances in our civilization. The first to see huge, distant bodies was Galileo, who in the seventeenth century devised an improved way to put lenses together to greatly magnify the heavenly planets and stars (fig. 1). Following this discovery of great importance and consequence, there has been a continuous improvement in telescopes to explore the sky. Today we have the Hubble (and similar) telescopes carried by a satellite orbiting the earth which not only has an enormously powerful telescope, but it can operate free of the optical disturbances created by the earth's atmosphere. The universe is unimaginably large, yet with this tool we are

Figure 1. Galileo's telescope. (Drawing by Hannah Bonner)

learning things about it that are totally beyond the reach of unaided human eyes.

The microworld is too small for us to see without help. That help first came in the seventeenth century when Anton van Leeuwenhoek ground very small lenses to greatly magnify objects (fig. 2). He not only invented the microscope, but he went on to illuminate a whole world that had never before been known to exist. He described for the first time all sorts of microscopic animals and plants that live in ponds and rain barrels: he even, for the first time, described bacteria, which are exceedingly small. He was the discoverer of spermatozoa in human semen, although it was not until much later that their nature was correctly understood. The technical advances in microscopes over the years has been no less remarkable than those of telescopes; their complexity and their power bears little resemblance to the tiny lenses ground out by van Leeuwenhoek many years ago.

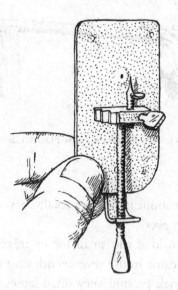

Figure 2. Van Leeuwenhoek's microscope. (Drawing by Hannah Bonner)

It has often struck me that, conscious as we are of our own size and of those around us, we do not ordinarily think that once we were microscopic. It is possible to believe that at one time we were infants, and there are all those old family photographs to prove it. It is even possible to imagine that earlier we were a foetus, but the idea that we were once a microscopic, single cell—a fertilized egg—is not something that ordinarily crosses our mind. In our own growth we go from the world of van Leeuwenhoek to the world we see around us every day. Perhaps if we were to have eyes and senses and a memory at that single cell stage of our life—an absurd possibility—we would look at the world differently, for our sense of size does not come to us until we are a small child. (In any event, an egg does not have much to look at in the darkness of a Fallopian tube.) One of my early recollections is being so small that I could not reach a light switch on the wall and had to get help from some giant adult. I had a psychologist friend who wanted to apply for a grant to build a huge room with gigantic furniture exactly as a small child would see it, and study the effect it might have on an adult who reenters his childhood world. The university committee that had to pass on research done on human subjects turned the request down: the effect might be too dangerous for the human psyche!

Clearly children are especially conscious of their size as they are surrounded by huge adults. This is reflected in fairy stories and children's tales. A good example is Lewis Carroll's *Alice in Wonderland*, where Alice keeps changing her size to go down rabbit burrows and back to her normal large self. She begins with the bottle which says "Drink Me" and shrivels to a size that allows her to join the underground world. Later, the hookah-smoking caterpillar puts her on to a magic mushroom: a bite from one side makes her larger, and a bite from the other smaller. What a perfect child's dream to go back and forth from one size world to another. At one point she worries that she will shrink to nothing, which has special resonance for me. When I was a very small boy I had a terrifying recurrent nightmare that I would enter a room with adults and I would instantly shrink to something minute, like an untied balloon letting out its air. I never knew my final size—I always woke up shrieking. Perhaps that was the first spark of an early interest in the role of size in biology, a subject that has pursued me these many years.

Another way in which we show our fascination with size is our sustained interest in human dwarfs and giants. In the nineteenth century, Barnum had great success in his circus exhibiting the midget Tom Thumb along with the giant Angus MacAskill, who were so extraordinarily small and large compared to everyday human beings. MacAskill came from Cape Breton in Nova Scotia, now an hour's drive from where I write these words. He was unusual in that he was not only huge—he was 7 feet and 9 inches tall—but he was phenomenally strong. His palms were almost eight inches across and he could hold Tom Thumb, who was a mere 3 feet tall, in his hand.

Both Tom Thumb and Angus MacAskill suffered from a malfunctioning pituitary gland: in Tom Thumb's case there was a deficiency in the secretion of growth hormone, and in MacAskill's case an oversecretion. It is a powerful hormone that controls our size during our growing years, and if it is not tuned just right, as it normally is, a dwarf or a giant will be the result. Today if the pituitary malfunctions, abnormal size can be prevented by controlling the amount of growth hormone. In the case of pituitary giants it is rarely the case that the individual will have the strength of MacAskill, and in general sufferers of the disease do not live to an old age. (MacAskill died at the age of 38 of "brain fever," presumably meningitis.) The excess of growth occurs mostly by stimulating the cartilage at the ends of the long bones, so it is possible to produce gigantism only in a growing youth. If too much hormone is produced after the long bones become sealed off and their layer of cartilage is gone, which usually occurs around the age of 20, only those parts of the human anatomy that still have cartilage will have a growth spurt. This will happen mainly in the hand and the face—a disease called acromegaly—and those unfortunates stricken with it will have enlarged jaws, broadened noses, and distorted finger joints. The cause is usually a tumor on the pituitary gland.

The fascination with giants and dwarfs goes way back in literature. Children's stories are replete with them, from Jack the Giant Killer to the Seven Dwarfs and innumerable others. The wonderful Icelandic sagas have diminutive, evil trolls as well as heroes of huge proportions who pitilessly slice off parts of the enemy with their gigantic broad swords.

Perhaps one of the best known and most enduring tales is *Gulliver's Travels* by Jonathan Swift. Lemuel Gulliver was about twelve times taller than the Lilliputians, and the

Brobdingnagians were that amount again taller than he was. Swift sensed some of the problems, and in order to feed Gulliver, the Lilliputians cubed their difference in height and for their half pint of wine they gave him 1,728 (12^3) half pints. We know today that this calculation does not take into account the relative differences in metabolic rate, and it would have made Gulliver roaring drunk, but Swift was taking a step in the right direction. In other things he made pure fictions, but interesting ones. According to Swift, being smaller meant being able to see smaller things; everything in the natural world of Lilliput was to scale—not just the people. The horses, the chickens, the sheep, the trees, the corn in the fields were all miniature so that their world appeared to them as ours does to us. From this perspective, the Lilliputians' ability to see small things was far greater than ours. "I have been much pleased with observing a Cook pulling a Lark, which is not so large as a common Fly, and a young Girl threading an invisible Needle with invisible Silk." A Lilliputian microbiologist would not need much of a microscope to help her see microbes. The corresponding difference is seen between Gulliver and the Brobdingnagians; he viewed their huge bodies as though through a magnifying lens. He watched the gigantic queen's maids of honor undress, "before their naked Bodies; which, I am sure, to me was far from being a tempting Sight, or from giving me any other Motions than those of Horror and Disgust. Their skins appeared so coarse and uneven, so variously coloured when I saw them near, with a Mole here and there as broad as a Trencher, and Hairs hanging from it thicker than Pack-threads, to say nothing further concerning the rest of their persons."

As we shall see, none of this is physically possible; no matter how big or how small a living object, the same physical laws will govern it, and those laws will make for differences in shape and form with differences in size. So the three size levels in Gulliver's world, with man and all the beasts being of similar proportions no matter their size, is impossible; it is the world of fantasy. Yet Gulliver is correct when he says, "Undoubtedly Philosophers are in the Right when they tell us, nothing is great or little except by Comparison."

THE SIZES OF LIVING THINGS

My plan is first to give the reader some feeling for the enormous variety of animals and plants and all the lower forms. Then in later chapters I will go on to examine what these size differences mean for all aspects of their existence: for their shape, for their complexity and their division of labor, for their evolution, for their abundance in nature, and for all those time-related activities such as metabolic rates, generation times, speed of movement, and even the pitch of their voices and songs.

Big animals and plants have a special fascination, beginning for us all in our childhood. My first biology book was a huge volume by H. G. Wells, Julian Huxley, and G. P. Wells called *The Science of Life* (1931). It was exciting reading for a small boy, and in fact it trapped me into becoming a biologist. In particular I was fascinated by a section on the size of organisms, which no doubt psychologically imprinted me so that after all these years I am writing this book! I still have my battered copy, and to this day I find the figures on the extreme size of a variety of plants and animals engrossing. They have a rather antique appearance, but they send a clear message.

Consider their figure of the largest forms (fig. 3). To make it easy to understand relative sizes, a man is included alongside the largest of all animals, a blue whale (1). As Galileo understood, it is to be expected that the champion would be an aquatic beast because the whale's specific gravity is close to that of water, and therefore it is essentially weightless and need not have supporting limbs.

Well-known dinosaurs, such as *Tyrannosaurus* (2) and *Diplodocus* (3), were among the biggest terrestrial animals, but there were larger ones. In fact, it was known since the early part of the twentieth century—well before the book was published in 1931—that the biggest of them is not among the popular ones shown in the figure. The record giant is *Brachiosaurus*, first discovered in Colorado in 1900, but a complete skeleton was not unearthed until a dozen or so years later in what was then German East Africa, now Rwanda and Burundi. The reconstructed skeleton can be found in the Humboldt museum in Berlin—standing in imperial elegance

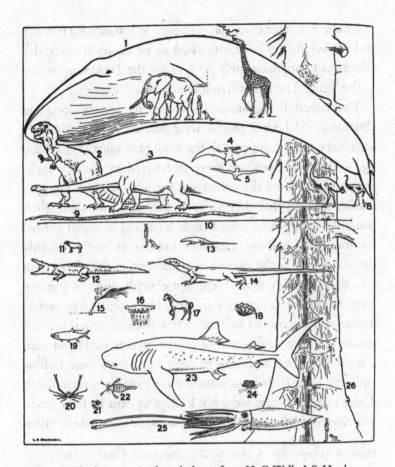

Figure 3. The largest animals and plants, from H. G. Wells, J. S. Huxley, and G. P. Wells, *The Science of Life* (1931). The names of the organisms are given in the text.

in a huge hall built especially to house it. It stands 40 feet tall, and its weight has been estimated to be about 80 tons; this compares to approximately 30 tons for the *Diplodocus* we see in the figure. The tallest living mammal is the giraffe.

The largest living animals that can fly are condors, and albatrosses (5) have a record wingspan of 11 feet, but they seem very small compared to the fossil pterosaurs (4) (see also fig. 4). Pterosaur remains were discovered in the late eighteenth century, and the celebrated French anatomist Georges Cuvier showed that they were reptiles and gave them their name. They appeared long before birds and bats and existed for about 215 million years, then became extinct at roughly the same time as the dinosaurs at the end of the Cretaceous era. Even though there were some small ones of birdlike proportions, the larger ones are truly huge (fig. 4). The record holder was discovered in Texas a few decades ago: it was estimated to have a wingspan of at least 36 feet, far bigger than a small airplane. The bones of these huge fliers were hollow and very light; their musculature must have been formidable. These record holders were not known to exist in 1931 when *The Science of Life* was published; if we were to add these giants

Figure 4 (facing page). Soarers. (a) The magnificent frigate bird has a wingspan of more than 6.5 feet; (b) the wandering albatross can have a wingspan of as much as 11 feet, the greatest of any living bird; (c) the largest flying animal of any age was the pterosaur, with a wingspan estimated at 36 to 39 feet. (From T. A. McMahon and J. T. Bonner, *On Size and Life*, 1983)

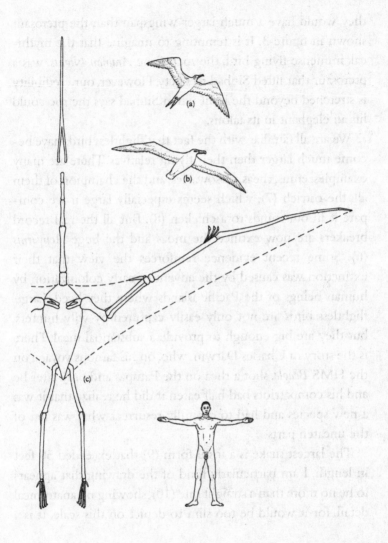

they would have a much larger wingspan than the pterosaur shown in figure 3. It is tempting to imagine that the mythical, immense flying bird, the roc of the *Arabian Nights*, was a pterosaur that lifted Sinbad to safety. However, our credibility is stretched beyond the limit when Sinbad says the roc could lift an elephant in its talons.

We are all familiar with the fact that flightless birds have become much larger then their flying relatives. There are many examples: emus, rheas, cassowaries, and the champion of them all, the ostrich (7), which seems especially large if we compare it in our minds to a chicken (8). But all the real record breakers are now extinct: the moas and the huge *Aepyornis* (6). Some recent evidence reinforces the view that their extinction was caused by the advent of early colonization by human beings of the Pacific islands where they lived. Large, flightless birds are not only easily captured by wily hunters, but they are big enough to provide a substantial meal. There is the story of Charles Darwin, who, on his famous voyage on the HMS *Beagle*, shot a rhea on the Pampas and only after he and his compatriots had half eaten it did he realize that it was a new species and had to carefully resurrect what was left of the uneaten parts.

The largest snake is a fossil form (9) that exceeded 50 feet in length. I am particularly fond of the drawing that appears to be no more than a straight line (10), showing no anatomical detail, for it would be too slim to depict on this scale. It is a record tapeworm found in a human being, and the fact it is so long means that our intestine is also long and highly convoluted.

Lizards can be very big, the largest being an extinct, fossil species (14), which is considerably bigger than a record crocodile from West Africa (12). The huge and ferocious Komodo dragon (13) is modest by comparison, but seen alongside a man or a sheep (11) it appears fairly large. Those who have seen nature films of these inhabitants of Komodo Island near Bali, and watched them tear a carcass apart in a viscous frenzy, do not think of them as being small; they are indeed dragons.

Among invertebrates the largest jellyfish is *Cyanea* (16), which was made famous in one of the Sherlock Holmes stories, *The Lion's Mane*, where it turned out to be the murderer. The biggest specimen known was found in the North Sea: its tentacles spread about 8 feet below the huge bell of jelly. No doubt it is a threat with all its innumerable tiny, poisonous nettle-like stinging cells; but so far as I know, Conan Doyle's fictional account is the only one in which it has actually killed a man. The largest related polyp is of comparable size (15). The giant clam (18) of the southern Pacific Ocean is very big compared to the ones we normally eat—all the invertebrates depicted here are in the same size bracket as a horse (17). Clams feed by bringing a current of water in between the valves and consuming the microscopic plankton they draw in; the valves are held together with a strong muscle that shuts them when the beast is disturbed. It has often been claimed that they can be accidental murderers if a diver inadvertently happens to get a foot inside the closing valves of a giant clam, but apparently it is no more than a tall tale, for no such lethal accident has ever been recorded.

To continue with the invertebrates, crustaceans can become quite large, such as the Japanese spider crab (20), which towers over a large Atlantic lobster (21). Even bigger is an extinct sea scorpion, or *Eurypterid* (22), a common fossil form. The biggest invertebrate of them all is a mollusk, the giant squid (25) that lives deep in the ocean and is a prey of the sperm whale.

Among fishes, a large tarpon (19) seems small alongside the huge whale shark (23), yet it in turn seems modest in size compared to a blue whale (1), which, of course, is a mammal.

The plant kingdom is less well represented in this figure, although this kingdom produces the biggest organisms of them all. A huge sequoia (26) is so big that only less than a third of it will fit on the page, and for comparison a 100-foot larch tree is superimposed. However, the size of trees is special in that a large part of their bulk consists of dead wood and the living tissue is a relatively thin layer under the bark, besides of course the leaves.

The largest flower is *Rafflesia* (24), which is found in Malaysia and Indonesia (fig. 5). Sir Thomas Raffles, the extraordinary British explorer and statesman who founded

Figure 5. The largest flower, *Rafflesia*. (Drawing by Hannah Bonner)

Singapore, discovered it in 1818. He encountered the flower in Sumatra and wrote in a letter,

> The most important discovery ... was a gigantic flower, of which I can hardly attempt to give anything like a just description. It is perhaps the largest and most magnificent flower in the world ... its dimensions will astonish you—it measured across from the extremities of the petals rather more than a yard ... and the weight of the whole flower fifteen pounds.

It is a parasitic plant that feeds off the roots of vines—it has no leaves. Not surprisingly, considering its weight, the flower lies on the ground, and it blooms for a brief but glorious four days. It has huge red petals that look like seat cushions, and produces a wonderfully offensive odor of putrefying flesh—irresistible to flies that pollinate the flowers.

Also not shown in the figure are some very large brown algae. These are the very big marine forms known as kelp. They have a holdfast that anchors them to the bottom of the ocean, and a long stem leads to the blades near the surface where they can bask in the sun for photosynthesis. There is an old account from one of the early sea voyages that a species of kelp (*Macrocystis*) can be over a 1000 feet long, and here lies an interesting tale: in the early part of the twentieth century, the longest measured was 138 feet, which is not inconsiderable.[1]

The older account of vastly exaggerated length died hard after the modern measurements were made in 1914. When I was a student, textbooks of elementary biology and botany still carried the tall tale even though quite a few years earlier it had been shown to be untrue. It was not until midcentury that the fiction began to fade, proving that we like large organisms, and if nature does not provide ones large enough we manufacture improvements.

A few years ago there was a big splash in the newspapers announcing that the biggest organism ever had been discovered.[2] It was a fungus that parasitized the roots of evergreen trees and spread huge distances in the forest, attacking and killing the trees as it advanced. It traveled over thousands of acres about 3 feet underground, and all the tips of the spreading fan of fungal filaments were shown to be part of the same individual; all were genetically identical. This is, however, a spurious individual organism; in no way does it qualify as a record giant organism. Soil fungi start from a germinated spore at one point and spread outward. It is commonly observed that they give rise to "fairy rings," a circle of mushrooms that form along the edge of the spreading filaments. On aerial photos of the Salisbury Plain in England they can be seen as discolored circles, and knowing how far they grow in a year one can estimate that some are 500 years old. The parasitic fungus mentioned above spread much farther and is estimated to be well over a 1000 years old. The difficulty is that the peripheral filaments become separated from their common stem—they are clones, not one giant beast. They are like groves of aspen (and some other plants) that spread by runners, but we do not consider the entire grove one individual.

Wells, Huxley, and Wells also examined the other end of the scale and show a rather old-fashioned drawing of a variety of small organisms that surrounded us: the microscopic world that van Leeuwenhoek opened up with his microscope (fig. 6). In a pond or in a swamp we are likely to find a number of protozoa that are single-cell animals. Mostly they are free swimming, although sometimes they are attached to the bottom, as in *Vorticella* (8). The one we remember from our

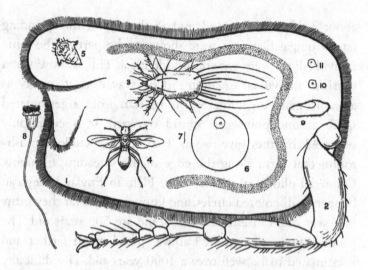

Figure 6. Some smaller organisms, from H. G. Wells, J. S. Huxley, and G. P. Wells, *The Science of Life* (1931). The names of the organisms are given in the text.

elementary biology course is *Paramecium* (9), and the largest ciliate protozoan is *Bursaria* (1); note its huge nucleus that looks like a long piece of rope. Number (11) is a human liver cell, and (10) an amoeba that causes dysentery. A very small wasp is seen in (4), a species of rotifer, which is a very small multicellular animal, in (5), and a cheese mite in (6). A human sperm (7) seems small alongside a human egg (6). Also shown is the foreleg of a flea (2). If this figure were done today it would include the recently discovered smallest vertebrate, a fish in Australia that is about 7 millimeters long and lives for only two weeks.[3]

While all of these organisms are very small, they are by no means the smallest. The most abundant of all the micro-organisms are bacteria, and they would be about the size of the stipple dots in the figure. There are even smaller organisms, the so-called mycoplasmas. Unlike bacteria, they lack cell walls and are a fraction of the size of bacteria. At the other extreme is a very large and curious bacterium that is big enough to be seen with the naked eye and lives in the intestine of some fish.

It is interesting to note that at one point in the early history of our Earth bacteria were the only organisms, and they have not only failed to become extinct, but are found everywhere today in incredibly vast numbers—a genuine case of multi-zillions. They have even been found in the deepest trenches in the ocean floor, and most remarkably in the crevices of rocks in the deepest mines. They also exist in great abundance in the intestines of animals. It is estimated that we harbor in our own gut 10^{13} bacteria (ten trillion), which is considerably more than the number of cells in our bodies. It is mostly the big beasts, such as the dinosaurs, that die out; the smallest ones started off as a success and have remained so ever since.

CHAPTER 3

THE PHYSICS OF SIZE

WEIGHT-STRENGTH

Size and shape are inextricably connected. One of the most elementary ways we can show that size is the supreme arbiter and has a profound effect on shape is to consider the relation between weight and strength. Galileo first demonstrated the principle elegantly in his book *Dialogues concerning Two New Sciences*, which he wrote after he was tried and condemned by the court of the Inquisition in Rome for having committed heresy in supporting Copernicus's heliocentric view of the heavens—that our world was not the center of everything. He wrote the book while in house arrest in the palace of the archbishop of Sienna, but he was forbidden to publish anything in Italy. An opportunity arose when the Dutch publisher Louis Elzevir came to Italy in 1636 and managed to take with him, on his return home, the then completed part of the manuscript. The subjects, mechanics and motions, were ones that Galileo had been working and lecturing on for many years, and now it was brought together in what he (and others) considered his most important work.

The book was written in Italian, and there is an excellent translation in English.[4] The science is presented in the form of a dialogue between three men: Salvati, Sagredo, and Simplicio. Today it is so unlike the way we present science that it seems rather contrived, but the main points are made with the utmost clarity, often illustrated by diagrams and drawings. There are four books, or Days, and in Day Two Galileo expounds the basic relation between size and shape. His main point is that the weight of any body is a cubic function, that is, the cube of its linear dimension, be it height or width. Strength, on the other hand, is a function of the cross-section area, that is, the square of the linear dimensions. Let me give you the consequences of this difference in his own words, which he relays through Salvati:

> From what has been already demonstrated, you can plainly see the impossibility of increasing the size of structures to vast dimensions either in art or in nature; likewise the impossibility of building ships, palaces, or temples of enormous size in such a way that their oars, yards, beams, iron-bolts, and, in short, all their other parts will hold together; nor can nature produce trees of extraordinary size because the branches would break under their own weight; so also it would be impossible to build

up the bony structures of men, horses, or other animals so as to hold together and perform their normal functions if these animals were to be increased enormously in height; for this increase in height can be accomplished only be employing a material which is harder and stronger than usual, or by enlarging the size of the bones, thus changing their shape until the form and appearance of the animal suggests a monstrosity. (Fig. 7.)

Salvati then goes on to say that because of this weight-strength relationship, a smaller animal will have a greater relative strength, which he illustrates by pointing out that "a small dog could probably carry on his back two or three dogs of his own size, but I believe that a horse could not carry even one of his own size."

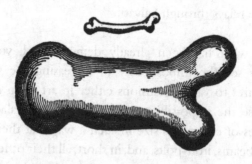

Figure 7. A drawing From Galileo's *Dialogues concerning Two New Sciences*, showing how a normal bone (top) would need to be increased in thickness (bottom) if the weight of the animal were much greater.

Simplicio then says he doubts this statement is true because of "the enormous size of certain fish, such as the whale." He was a bit off calling a whale a fish, but Salvati has a ready answer to this welcome question: by being submerged in the water, a whale is "making their bones and muscles not merely light but entirely devoid of weight."

The consequences of the weight-strength ratio being influenced by size are large, and here I will stress the living world rather than that of the engineer, although both are equally affected. Beginning with the animal world, the fact that a small gazelle has slender, graceful legs, and an elephant is endowed with great stumps is precisely what Galileo would have expected. If there were elephants on the moon, which is smaller than the earth and therefore the gravity is less, one could imagine that they would have evolved into gigantic gazelle-like beasts. Yet another way to look at the same point may be found in that splendid chapter in D'Arcy Thompson's book *On Growth and Form*. He points out that the size of a terrestrial vertebrate is reflected in the relative weight of its bones, and consequently, "bones make up some 8 per cent of the body of a mouse or wren, 13 or 14 per cent of goose or dog, and 17 or 18 per cent of the body of a man."

As Galileo makes clear, the shape of trees is equally affected by size, and we expect a giant sequoia to have, like the elephant's foot, a huge wide and solid base compared to a smaller tree. The difference can even be seen in the growth of a single tree: a maple starts off as a slender, willowy sapling that, with increasing size, becomes disproportionately thick in its trunk and limbs. The same principle can be seen in fruit of different sizes. The strength of the stem of each fruit will depend on its cross-section area, so the stem of an apple will be disproportionately thick compared to that of a cherry. The stems of coconuts are very thick and sturdy, while watermelons have solved the problem by lying on the ground; their solution is somewhat analogous to that of whales, for in different ways both have largely circumvented the pull of gravity. The relationship between strength and weight can be expressed in a simple equation (to which we will return):

$$(1) \text{ Strength} \propto \text{Weight}^{2/3}$$
$$(\propto = \text{varies as}).$$

BACK TO GULLIVER

To return to *Gulliver's Travels*, let us reexamine Jonathan Swift's work in the light of our discussion; it is a good way of illustrating the principles laid out here.

Let us assume that Lemuel Gulliver was 5 feet and 8 inches tall. The Lilliputians were one-twelfth his height, which would make them just shy of 6 inches tall—roughly the size of a mouse. This would mean that their legs and their bones could be relatively slender compared to normal human beings; they could have stick legs like a sandpiper that runs along the edge of the water on the beach. After all, if Gulliver weighed about 150 pounds, then the similarly constructed Lilliputian would weigh less than a pound. The result would be that he would not be similarly proportioned to Gulliver, for legs like match sticks would be quite sufficient to support him.

The speed of running of Lilliputians, while no doubt slower than that of Gulliver, would not be that far behind because their legs will move much faster (like the sandpiper), and for one of Gulliver's normal steps they would have taken many. All the movements of the smaller homunculus would be more rapid than our movements, just as the wing beat of a small hummingbird is far quicker than that of a swan.

We know that rats and shore birds have lungs, a heart, gut, and circulatory system as we do. The only difference is that the proportions of these parts of their anatomy will differ from ours, for they have relatively less work to do because of their smaller size: the heart pumps the blood into a smaller set of blood vessels and therefore can be smaller. Because the Lilliputians have smaller vocal chords their voices will be high and squeaky, and they would talk with great rapidity, so much so that it might be difficult for Gulliver to hear and understand them.

The reverse of these arguments would apply to the huge Brobdingnagians, who, to begin with, would have extremely deep and slow voices. Swift describes the first of the giants he encountered this way:

> He appeared as Tall as an ordinary Spire-steeple; and took about ten Yards in each Stride, as near as I could guess. I was struck with the utmost Fear and Astonishment, and ran to hide my self in the Corn, from whence I saw him at the Top of the stile, looking back to the next Field on the right Hand; and heard him in a Voice many Degrees louder than a speaking Trumpet; but the Noise was so High in the Air, that at first I certainly thought it was Thunder.

If their proportions are those of a normal human being and they are twelve times taller than we are, which would make them 68 feet tall, they would weigh between 12 and 13 tons. This makes them lighter than the larger dinosaurs, but about ten times heavier than a large elephant. Like these large beasts, the Brobdingnagians would require greatly thickened legs and leg bones, so much so that they would no longer appear like normal human beings, but more like victims of advanced elephantiasis of the legs.

SIZE AND DIFFUSION

Strength is not the only feature that affects the shape of an organism as its size changes. Another most important one is diffusion, which, as we shall see, has a powerful effect on shape.

Diffusion is always through surfaces, and since the surface is a square of the linear dimensions and the weight is the cube, we can again write a simple equation to express the relation of surface to size:

$$(2)\ \text{Surface} \propto \text{Weight}^{2/3}.$$

Most organisms, including ourselves, depend on oxygen for their existence; every cell of our body needs to have oxygen diffuse into it in order for it to function and stay alive. Diffusion is the random movement of molecules from regions of high concentration to low, so that ultimately, at equilibrium, they are evenly spread out. Molecules zip around at great speed when they collide with one another. In regions where they are numerous, they collide often and have more luck spreading when there are fewer ones to bump into. A way to look at this might be to imagine a tea bag placed in a glass of cold water and all currents, convection and otherwise, are eliminated. The brown color released from the tea bag is first concentrated in and around the bag itself, but eventually the brown color will be evenly distributed; it has diffused from the bag.

If oxygen has to get into an animal or to its cells it must diffuse through a surface that, like strength, is a function of the square of the linear dimensions. This can be illustrated in a simple fashion. Consider two hypothetical spherical organisms, one the size of a sand grain a millimeter in diameter, and the other the size of a golf ball. Oxygen is a small molecule that diffuses rapidly, and in a matter of seconds it can penetrate a millimeter into a living body. The middle of the tiny sphere is close enough to the surface so that as soon as it uses up its oxygen it will be replenished by a steady diffusion of more from the surface. But in the golf-ball beast it would take an hour or so to reach the interior, and we all know that would not do—if we are deprived of oxygen we die. In fact, we are so tuned in to this problem we cannot even hold our breath for more than a few minutes; we have a built-in mechanism that forces us to start breathing again.

The only way an organism larger than a millimeter can survive is to bring its interior near the surface everywhere so that no part is farther away than one millimeter. This can be managed by radically increasing the amount of surface, making many convolutions and folds so that every point in the interior is close to the surface. Instead of a perfect, smooth sphere, we now have a golf ball with an incredibly wrinkled surface (fig. 8). Or to put it another way: because the enlarged spherical organism cannot exist in it original form, it must grow more skin, and the result will be a radical change in shape.

Shape changes caused by diffusion problems can be far more revolutionary than this transformation. Much of the shape of human beings is molded by the diffusion of oxygen and the problem of getting it to our inner cells. To achieve this end, we have a marvelously complex set of devices. We have lungs that bring the oxygen close to our blood vessels

Figure 8. The shape of a hypothetical golf-ball-size organism that requires every internal cell to be near the surface. (Drawing by Hannah Bonner)

where the hemoglobin can pick it up; we have a circulatory system with a heart, a most ingenious and durable pump that pushes the oxygen-containing blood into the minute vessels, or capillaries. They permeate all parts of our bodies and lie close—well within the one millimeter limit—to each of our deep-seated cells. Note that the lungs and the blood vessels have vast surfaces; like our wrinkled golf ball, they keep the surface-volume ratio constant despite our large size. More than that, they have set up a system of motion so that the lungs move the air in and out, and the beating heart keeps the blood circulating. The movements supplement diffusion so that fresh oxygen can be rapidly picked up in the lungs and delivered to the cells.

Oxygen is not the only bit of fuel needed to keep a large animal going. Food to burn for energy is needed as well, and it also must be taken in through a surface. Our food canal, from our mouth to our anus, is a tube, a surface through which the broken-down, dissolved food has to pass. The greater the surface of our gut, the greater our ability to absorb food. Because surface increases as the square, the larger the size of an animal (which increases as the cube), the greater must be the surface of the gut to keep pace in order to maintain a constant surface-volume ratio. This change can be seen clearly if we compare ourselves to minute nematode worms that are little bigger than an eyelash. Their gut is a straight tube, but ours is a tortuous, highly convoluted tube many times our length. Furthermore, our small intestine is not a simple tube, but has a forest of tiny papillae that stick up into its inside, each one greatly increasing the gut surface to absorb more food. We have also developed glands: the salivary glands, the pancreas, and the liver that produce batteries of enzymes to break down the food into small molecules such as simple sugars and amino acids, so that they can diffuse through the gut wall more easily.

All these devices found in a large animal are present to get energy from food by combusting it with oxygen in all the cells of the body. This job calls for lungs, a circulatory system, and a highly elaborate gut and its associated glands. From this necessity it is clear that size increase requires quite extraordinary changes in shape. In no way do we resemble a smoothly spherical organism, nor even a wrinkled golf ball; we have evolved into a far more complex shape largely because of the increase in our size.

Viewed through the eyes of Lemuel Gulliver, the Lilliputians need only a comparatively small number of blood vessels to reach all the tissues of their bodies. The Brobdingnagians, on the other hand, need a vast number, and their guts will have to be disproportionately extended with an immense increase in their surface area to absorb sufficient food to accommodate their giant size.

SMALL WORLDS

It is possible to put large and minute organisms in quite separate baskets; there is a fundamental rift between the living worlds of the large and very small, even though it is obviously a continuum if one goes from large animals and plants to small ones; organisms come in all sizes in between. The reason for arguing for such an artificial separation of two size worlds has to do with the forces affecting the totally different size levels. The effect of gravity becomes progressively insignificant the smaller an organism, while molecular forces come to the fore and play an increasingly important role. Even though these forces decline and rise smoothly as the size changes, one can artificially designate a critical point where one jumps from one world into the other. Presently I will give an example, but first let us look at a simple graph (which I have taken from a splendid paper by F. W. Went) in which the force of gravity and the force of molecular cohesion are plotted on a size scale (fig. 9).[5] In terrestrial organisms, gravity varies with the weight, and therefore with the cube of the linear dimensions, while the force of attraction between surface molecules varies roughly as the linear dimension. This approximation takes into account the forces of attraction between molecules and the distance between the molecules on the two surfaces. The important point is that the cohesion curve is less steep than the curve for gravity. The figure is drawn in such a way that the lines for gravity and cohesion arbitrarily intersect for an organism roughly a millimeter in length; above that

Figure 9 (facing page). A synthetic graph comparing how different size terrestrial organisms are affected by gravity and cohesive forces. Forces due to weight vary as the length cubed, while the forces due to molecular cohesion average out very roughly to be proportional to the length. The lines are arbitrarily placed so that they cross for organisms 1 mm in length, and it can be seen that in larger organisms the forces of gravity (which vary with the mass) play a far more significant role than the forces of cohesion, while for smaller organisms cohesion forces dominate. (This figure is based on one in F. W. Went, *American Scientist* 56 [1968]: 400–413)

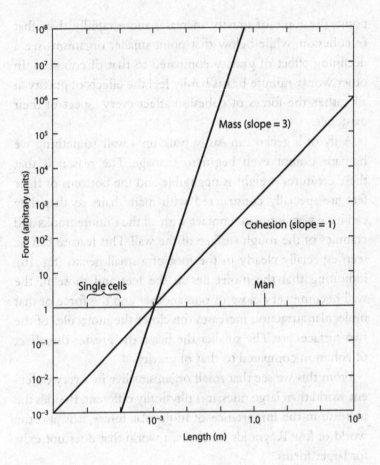

point, the force of gravity increases more rapidly than that of cohesion, while below that point smaller organisms see a declining effect of gravity compared to that of cohesion. In other words, minute beasts hardly feel the effects of gravity at all; rather, the forces of cohesion affect every aspect of their existence.

A fly or a gecko can easily walk up a wall, something we humans cannot even begin to manage. The reason is that these creatures' weight is negligible and the bottoms of their feet are specially constructed with many hairs so that they can have very intimate contact with all the minute nooks and crannies of the rough surface of the wall. This feature can be seen especially clearly in the foot of a small gecko (fig. 10), indicating that the molecules in the feet and those in the wall become very close to one another and the force of that molecular attraction increases the closer the molecules of the two surfaces are. The smaller the beast, the greater the effect of cohesion compared to that of gravity.

From this we see that small organisms live in a very different world than large ones; it is physically different. Besides the increase in the importance of molecular forces, it is also the world of low Reynolds numbers, a world that does not exist for larger forms.

The Reynolds number is a convenient way of describing the immediate environment of a moving body. It is a way of showing that a swimming bacterium is in a very different environment than a swimming whale even though both are

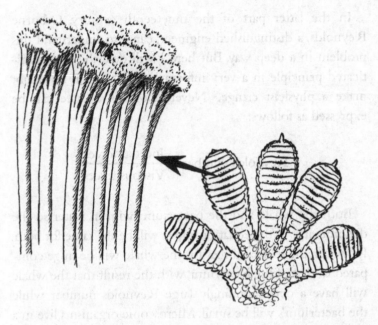

Figure 10. The foot of a gecko and a high-power microscopic view showing the mass of very fine hairs that can make intimate contact with a solid surface. (Drawing by Hannah Bonner)

swimming through water. The vast difference in their size, for purely physical reasons, affects their locomotion. A whale has a far greater inertial force, and if it suddenly stops swimming it will glide on for some time. Because of its great mass it is endowed with a large amount of momentum. On the other hand, when a minute bacterium stops swimming, its mass is so small that it stops in a split second.

In the latter part of the nineteenth century, Osborne Reynolds, a distinguished engineer, came to understand this problem in a deep way. But here I will describe his sophisticated principle in a very informal way that will no doubt make a physicist cringe.[6] Nevertheless, its essence can be expressed as follows:

$$\text{Reynolds number} \propto \frac{\text{Inertial forces}}{\text{Viscous forces}}.$$

Both the whale and the bacterium swim in water so the denominator in the equation above will be the same for both. However, the inertial forces for the whale will be huge compared to that of the bacterium, with the result that the whale will have a correspondingly large Reynolds number while the bacterium's will be small. Microscopic organisms live in a world of low Reynolds numbers.

To give one a feel of what this means for minuscule organisms, the noted physicist Edward Purcell painted the following picture.[7] Imagine a man swimming in a liquid that produces the same low Reynolds number as would exist for his own sperm. It would be like trying to swim in thick molasses in which one was not allowed to move one's arms or legs faster than the hands of a clock. Under these rules, if one moved a couple of meters in a couple of weeks, one would qualify as a low Reynolds number swimmer. This vivid picture gives one an idea of how the changes in the inertial forces affect a bacterium; it is indeed a different world. Purcell says, "Motion at low Reynolds numbers is very slow, majestic and regular."

Swimming microbes have whiplike structures called cilia or flagella that propel the organisms through the water. In single-cell protozoa, cilia move the organism by behaving like sophisticated oars; they are flexible and bend on the back stroke, but are stiff when pushing forward and provide the cell with a powerful thrust (fig. 11a).

Let me interject an aside here. When I was a student and first saw, through the microscope, these little whiplike structures move one-cell organisms across the field of vision, I marveled at how fast they traveled. Later I realized that they were in fact moving very slowly and the magnification of the microscope had fooled me. After all, speed is distance divided by time (as in miles per hour), and the high power of the microscope made the minute distance appear very large. For me it was part of growing up: I had not realized that what I was seeing—those protozoa zipping around at an incredible rate—is an optical illusion created by the magnification of the microscope. It was rather disappointing to learn that the microworld was not dazzlingly fast but rather sedate.

Bacteria are very much smaller and for that reason have a different set of problems. Not only are they in an environment of even lower Reynolds numbers, but also they are so

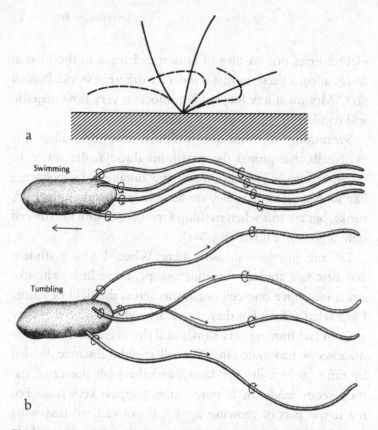

Figure 11. Microbe propulsion. (a) The flexible oar of ciliate. The solid curves show successive positions of the cilium beating downwards for forward movement; the broken curves show the cilium on the reverse stroke (from T. A. McMahon and J. T. Bonner, *On Size and Life*, 1983). (b) A bacterium (*E. coli*) swimming forward (top); and how a sudden reversal of rotation causes the flagella to splay out and the cell to tumble (bottom). (After J. Adler and H. C. Berg; from Bonner, *The Evolution of Culture in Animals*, 1980)

small they cannot construct the relatively huge and complex oarlike cilia or flagella of larger microbes. They are in the nanoworld, having very thin flagella of a simple construction that produces locomotion in quite a different way. Each flagellum is attached to a rotor that goes around and around, causing the flagella themselves to rotate like a ship's propeller, and this action thrusts the bacterium forward (fig. 11b). There are many remarkable aspects of this ingenious method of dealing with the difficulty of moving a minute body in conditions of very low Reynolds number.

The discovery that bacteria move this way was made not that many years ago. It was first predicted on theoretical arguments based exactly on what we have just seen: the special problems connected with low Reynolds numbers. This prediction was followed by a very clever experiment in which the flagella were trapped on to a specially prepared sticky microscope slide with the amazing result that the bodies of the bacteria could be seen, through the microscope, to rotate continuously. They possessed a true rotor, something not previously known to exist in the living world.

We often congratulate ourselves on our inventiveness that places us above all other animals. We not only discovered how to control fire and make use of it, but we also invented the wheel, that clever device that allows us to move heavy loads. It is usually claimed that no other animal can manage this feat, but as we now realize, it was in fact devised billions of years before *Homo sapiens* appeared on the globe. As we have just seen, bacteria invented the wheel, and exploited theirs in a most sophisticated manner.

To illustrate this sophistication, let me briefly explain how bacteria can orient themselves in their aqueous environment. It is well known that swimming bacteria will go towards food and away from noxious substances; now we know how they do it. They have molecules on their surface that can adhere to specific external chemicals, and if they happen to be swimming in the direction towards food, then there is a molecular signaling mechanism that tells the rotor to keep going. If, on the other hand, the food molecules become increasingly less abundant because they are swimming away from the food, then the rotor gets a new message and abruptly reverses its direction. This action causes the whole cell to stop, to tumble, and when it starts up again there is a good chance it will be pointed in a new direction, which, with luck, might be the direction of the energy-rich food. So bacteria not only can move, but they are blessed with the ability to move in a direction that is to their advantage. And they have managed all this without a brain!

A low Reynolds number environment also influences very small flying insects. Their force of inertia is insignificant, with the result that air is also like molasses. As a consequence in more than one group of insects, their wings have become transformed into what look like minute feather dusters

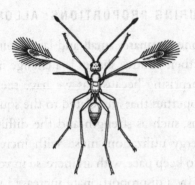

Figure 12. Among the smallest insects are fairy flies, which are actually tiny wasps. Their body lengths can be less than half a millimeter. The wings are slender stalks, whose margins are fringed with hairs. (From T. A. McMahon and J. T. Bonner, *On Size and Life*, 1983)

(fig. 12). These so-called fairy flies (which are in fact minute parasitic wasps) use their wings the way ciliate protozoa use their cilia; they swim through the air.

It is clear that size, from the pure point of view of physics, has an enormous effect on living organisms. It constrains, or even controls, the shape of an organism, including its internal structure. It determines what is possible and what is impossible. And as we have also seen, it can affect the behavior of animals as well as their locomotion. In the next chapter we will see that size is also directly related to the degree of division of labor.

MEASURING PROPORTIONS: ALLOMETRY

Note that if one compares small and large animals or plants, one aspect of their relative shapes is a change in proportions. This is not surprising because, as we have seen, in order to exist those properties that are related to the square of the linear dimensions, such as strength and the diffusion processes related to energy utilization, must, with increased size, rise more rapidly to keep pace with an increase in volume or mass. It shows itself by a disproportionate increase in the thickness of limbs and a more convoluted gut in larger animals.

There is a simple and elegant way to describe the changes in proportion to size, known as allometry.[8] Consider our earlier example of the bones of small gazelles and large elephants. If they were similar in all respects, then an elephant would have legs that looked like a photographic enlargement of a gazelle. If this were so and we plotted the height of animals of different sizes against the diameter of their bones, we would get a straight line with a slope of 1 (fig. 13). That is, the diameter varies as ($\propto$) the height:

$$\text{Diameter} \propto \text{Height}.$$

As we learned from Galileo, they are not similar because the overall size (weight) of the animal rises as the cube of the linear dimensions, while the strength of the bones rises as the square of the linear dimensions. That is,

$$(2)\ \text{Strength} \propto \text{Weight}^{2/3}.$$

This equation neatly expresses how the strength-weight affects the proportions. It is convenient to plot this relationship on the graph as logarithms, for the exponent of 2/3 will produce a straight line whose slope is 2/3 (fig. 13).

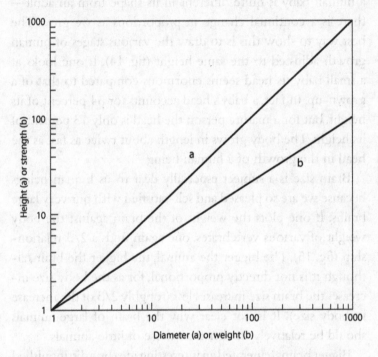

Figure 13. A log-log graph showing allometric relations. (Line a) If diameter is plotted against height in geometrically similar organisms of different sizes, then the slope of the line equals 1. (Line b) If strength (which is diameter squared) is plotted against weight, which is the cube of the linear dimensions, the slope of the line equals 2/3.

There are many ways to illustrate the usefulness of allometry, and I shall pick two: size changes during human growth, and relative brain sizes.

Proportions change as animals and plants grow. For instance, a human baby is quite different in its shape from an adult—there is a continual change in proportions as we grow. The best way to show this is to draw the various stages of human growth adjusted to the same height (fig. 14). If one looks at a small baby its head seems enormous compared to that of a grown-up. In fact, a baby's head accounts for 34 percent of its height, but for a mature person the head is only 13 percent of its height. The body grows in length about twice as fast as the head in the growth of a human being.

Brain size is a subject especially dear to us human beings because we are so pleased and self-satisfied with our very large brains. If one plots the weight of the brain against the body weight of various vertebrates, one again finds a 2/3 relationship (fig. 15). The bigger the animal, the bigger the brain, although it is not directly proportional, for as the body size increases the brain size increases less (roughly 2/3 of the increase in body size). It is not clear why the brain of large animals should be relatively smaller than those of little animals.

Bigger brains generated an interesting idea by a distinguished German evolutionary biologist named Bernard Rensch some years ago.[9] Rensch argued that being bigger meant having a larger brain, which meant that larger animals must be smarter.

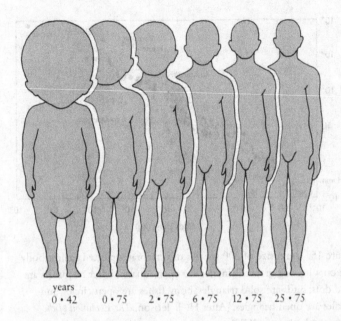

years
0 • 42   0 • 75   2 • 75   6 • 75   12 • 75   25 • 75

Figure 14. Changes in body proportions in a human being with increasing age. (After Wells, Huxley, and Wells, 1931; from T. A. McMahon and J. T. Bonner, *On Size and Life*, 1983)

Increased size automatically conferred such an advantage, and this was one of the reasons there are often evolutionary trends towards larger size. Rensch used the ability to learn, and retention of what is learned in the form of memory, as his measure of intelligence. To examine his idea he compared similar birds of different sizes—bantam chickens,

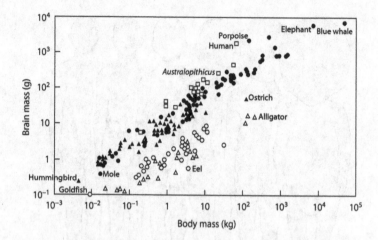

Figure 15. Brain size of 200 species of vertebrates plotted against body size on a log-log graph. Primates are open squares; other mammals are solid dots; birds are solid triangles; bony fishes are open circles; and reptiles are open triangles. (After H. J. Jerison, *The Evolution of the Brain and Intelligence*, 1973)

normal-sized ones, and giant breeds. He gave them a simple learning procedure followed by periodic testing to see how long they remembered what they had learned. He found that the larger birds could learn more and retained the newly acquired information for a longer period of time. Having a big brain did seem to provide an advantage. He realized that all of this fits in with the old adage that an elephant never forgets, and proceeded to make some interesting observations on circus elephants, and indeed they had a remarkably long memory for some card puzzles he taught them. I am reminded of a story my father used to tell us children about the circus elephant that recognized his old trainer in the audience and lifted him up with his trunk and moved him from the fifty-cent seats to the dollar seats.

Rensch once came to our university to lecture on all this, and in the discussion period I asked him if perhaps it was not intelligence, but because the animal's metabolism is slower the larger the animal, the elephant was simply slow to forget. Professor Rensch got quite irritated with this fresh question from a young man, but to this day I wonder if I did not have a point. Later I plan to talk about size and metabolic rate in some detail. I cannot believe that Rensch's intriguing idea has been an important factor in evolution. There are too many small animals of remarkable intelligence, such as parrots and crows among numerous others. One cannot imagine that the size of elephants increased because of a selection advantage of being better at learning and remembering things.

If we turn to primates, we see a similar relationship: the larger the monkey or ape, the larger the brain. However, if we look to the evolution of human beings, from *Australopithicus* to *Homo*, then we see an amazingly steep rise of brain size relative to body size. Instead of the brain enlarging at a slower rate (2/3 or less) than the body size, it rises at a much greater rate—almost two times that of the increase in the body size (fig. 16). Note that if we compare *Australopithicus* species of

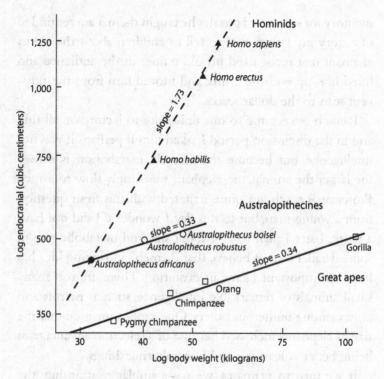

Figure 16. The endocranial volume plotted against the body weight of great apes, Australopithecines, and the *Homo* line on a logarithmic scale. For the fossil forms the weight is, of course, only an estimate. (Redrawn from D. Pilbeam and S. J. Gould, *Science* 186 [1974]: 892–901)

different sizes, the allometric relation of their brain size increase is modest, mirroring that of apes. In other words, in our evolution from a common ancestor with apes, one great difference from them is our remarkable increase in brain size.

We did indeed become the brainy animal. The only other animals that had a similar increase in relative brain size are porpoises and whales; we are only beginning to understand something of their complex behavior.

As one might expect, there has been a considerable amount of speculation as to why brain size might have made these sudden spurts. One of the earlier hypotheses centered around the idea that it had something to do with needed extra brain power to make and use tools. For various reasons this notion has fallen in disfavor, largely because tool making has come to light not only in great apes but in many animals. For instance, Caledonian crows will not only use a piece of wire to extract food from a long, narrow glass vessel, but after a few unsuccessful tries they will carefully bend the tip of the wire to hook the food at the bottom.[10] It seems unlikely that devising and using tools would drive brain evolution to any great degree; those Caledonian crows have the same size brain as other crows. Another idea is that the brain increase is associated with the rise of human language, which implies that in terms of Darwinian natural selection, being able to speak more effectively conferred a selective advantage, and the better communicators had more offspring to whom they passed on those skills. The difficulty with this suggestion is that an increase in brain size along with communication skills is merely a correlation; it is hard to see what comes first—what caused what.

A related and more general argument comes from the modern study of primate social groups, from lemurs to chimpanzees. Clearly, reproductive success in those groups does not just depend on brute strength, but on careful manipulation of one's position within the group. This in return requires managing alliances with others in the group to strengthen one's social position; it is what Frans de Waal has called, based on his fascinating studies, "chimpanzee politics."[11] This whole idea has now taken hold and has even been given a name: "Machiavellian intelligence."[12] According to this hypothesis, it is the social interactions that are key, and the individuals who can influence others to help them in their power struggle are the ones blessed with reproductive success. It is of special interest that recent work on the large-brain dolphins shows some strikingly elaborate social behavior; maybe they too have become big brained due to Machiavellian intelligence.

Because a large brain is such a striking character of *Homo sapiens*, great efforts have been put into attempts to tell us something about differences among humans. Using head or cranium size as a measure of brain size, there is great variation in our species. Those who have some sort of dubious agenda have especially loved this: those who scrounge for evidence to prove that a race with bigger heads is superior in intelligence to another with smaller ones. It has even been used to argue that men are superior to women because they have, on average, larger bodies and therefore larger heads!

Over and above the fact that this approach is distasteful, for many reasons it has proved to be a fruitless and muddled enterprise. To begin with, variation in brain size, as measured by volume within the cranium, very likely is too rough an approximation of the number of neurons in the brain and their pattern of interconnections. Variation in how closely packed the nerve cells are in different parts of the brain, and the degree of folding of the brain surface, will not be reflected in head size. In a splendid play about the famous mathematician Alan Turing, who did so much to launch computers and artificial intelligence, there is a scene where he goes back to his old school to give a lecture about the brain, which he describes to the boys as looking like a lump of porridge. In this chaos we know little about how the millions of neurons, and even more importantly how their interconnections, lead to intelligence. To make the point in a forceful way: thanks to the pioneering work of Karl von Frisch, we know that honeybees have an extraordinary ability to memorize information and even communicate that information to their fellow workers, yet the brain of a bee is minute with a correspondingly minute fraction of the neurons found in any mammal, let alone human beings. This is a fascinating area of inquiry, and despite recent considerable advances, we have yet to scratch the surface.

Furthermore, our crude methods of measuring human intelligence are woefully inadequate. IQ tests tell us something about thinking, but one never quite knows what. There is a strong suspicion among many, including myself, that they do not say anything about what we would normally recognize as intelligence, but merely tell us how good we are at taking that kind of a test. I have already pointed out that brain size varies with body size and therefore it is to be expected that women, who are on the average smaller than men, should have smaller brains, but there is no evidence for any significant difference in their mental abilities, even their ability in taking IQ tests.

As one might imagine, there is a great literature on this subject, often polemical. To my mind it can be brushed aside in a simple way. There are many well-known savants who had small skulls containing great minds. An example of such a small-headed person is Anatole France, the French author who is famous for his novels and for having exposed the anti-Semitic injustice of putting Dreyfus in the penal colony on Devil's Island at the turn of the twentieth century. In looking this matter up, I was delighted to discover that Jonathan Swift of *Gulliver's Travels* had an exceptionally large head—twice the size of Anatole France's!

FIVE RULES

Let us pause for a moment to comment on the two rules that have just been discussed. Both strength and surfaces vary with the weight in the same way:

$$(1)\ \text{Strength} \propto \text{Weight}^{2/3}$$

$$(2)\ \text{Surface} \propto \text{Weight}^{2/3}.$$

Later we will see that other important properties of living organisms vary in a similar way, although the exponents, and therefore the slopes of the lines when plotted logarithmically, vary. The three examples to come are

$$(3)\ \text{Complexity} \propto \text{Weight}^a$$

$$(4)\ \text{Abundance} \propto \text{Weight}^{-b}$$

$$(5)\ \text{Metabolic rate} \propto \text{Weight}^c.$$

All five reflect the fact that size differences, as indicated by their weight, affect different properties: for a given weight, an appropriate strength is required for support and movement; an appropriate surface, so that an adequate diffusion of oxygen and food substances can reach the inner tissues; an appropriate division of labor, so that the body of the animal or plant can function; and the appropriate rate of metabolism for that functioning. Finally, size affects the abundance of animal and plants in nature: the bigger the organism, the more space it needs. These five rules are central to the argument of this book; they are the pillars that underpin the effects of size on living organisms.

CHAPTER 4

THE EVOLUTION OF SIZE

SIZE INCREASE OVER GEOLOGICAL TIME

In looking at the history of the earth, there is general agreement that at some early era all organisms were unicellular; first bacterial (prokaryotic) cells, and later cells with a nucleus (eukaryotic). This was followed by the advent of multicellularity, and ultimately those multicellular animals and plants became the huge organisms I have just described. Over the years, from Aristotle on, some form of this trajectory was interpreted in terms of progress. Aristotle called it the scale of nature, which went from vegetables and worms to human beings. Because we were the pinnacle of this progress, it was not a perfect progression in size increase but rather one of inching towards the complexities and the great virtues of our own species.

Largely as the result of many discoveries of fossils, a similar idea sprang up in the nineteenth century. Fossils of some animals were well preserved, and it could clearly be shown that as geological time advanced the descendants of those animals became larger. One of the best known examples was the evolution of horses, where the dawn horse is the size of a fox, and with time the fossil remains showed increasingly large animals until we reach the large modern horse. There are similar fossil series for numerous other animals, such as the camel.

By noting these size-increase series over time, it was common, at least at the end of the nineteenth century, for people to consider that there was a mystical evolutionary force that governed these trends. The process was even given the fancy name of "orthogenesis"—it was an innate tendency in evolution for progress. The remarkable thing is that this continued many years after Charles Darwin's publication of his theory of natural selection that was quite incompatible with the idea of orthogenesis. Darwin pointed out that all species showed variation—inherited variation—and that the variants that were best adapted to a particular environment were the most successful ones in producing offspring. This was not a goal-directed process; there is no built-in mechanism that drives evolution in one direction; there is no "progress." The fact is that there were very few biologists until the 1930s who believed that Darwin's natural selection was the whole story; there had to be more, and for this they assumed some innate notion of progress, apparently undismayed by the fact that this was no explanation, but just a name for an unknown, assumed mechanism. Furthermore, it was not until much later into the twentieth century that animal series were discovered that decreased in size over geological time: size trends could go up or down.

That natural selection is all that is needed to explain the evolution of size increase from single-celled bacteria to blue whales and giant sequoias is taken for granted today. There is no longer a psychological need to assume unknown forces; we are comfortable with the simple and totally undirected mechanistic explanation. There has been a selection for larger organisms; under some circumstances size increase can confer advantages and promote reproductive success. This argument applies at the smallest level from unicellular to multicellular organisms. If there is an advantage to increased size, one of the simplest ways to achieve it is to become larger by becoming multicellular.

It is important at this point to understand the nature of natural selection for size; there is an ecological principle. If one looks at any environment, from fields to forests, from ponds to oceans, there is always an array of organisms of different sizes, from the smallest bacteria to the largest trees or vertebrates and all the middle-sized forms in between. The organisms at each size level are competing with one another for food, either by being more successful at catching or gobbling up the smaller prey if they are animals, or by being more successful at catching the sun's rays if they are photosynthetic plants. In a balanced environment, each size level will have certain advantages; in a sense they have their own house, their own right-sized niche that they can inhabit. If such a niche is vacated by an extinction, then it will soon be filled, for natural selection will favor the small individuals of a large species to become smaller, or larger individuals of a small species to become larger (or, of course another species of the same size). So each size level in the ecological environment will fill up if vacated. The process that achieves this is Darwinian natural selection.

If put in this way it is immediately evident that the size level that is always open is the largest—there is always room at the top. So if one follows the course of evolution over time, one would expect that the number of the size levels would steadily expand since the one at the top is always open and available to be filled. This simple fact explains why the world has evolved from one in which there was nothing larger than bacteria to today's, in which we have blue whales and giant sequoias. Let us examine this trajectory more carefully by looking at the fossil record to see if it conforms to the principle that there will be an expansion of the upper size limit throughout geological time. One can plot the largest organism known at any one time in earth history, and it is clear that this upper limit steadily increases from the beginning of life through today (fig. 17).

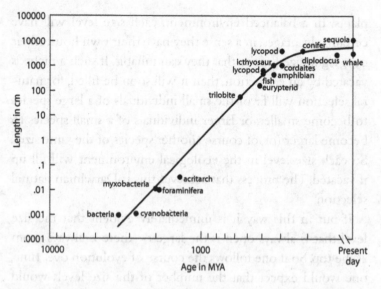

Figure 17. A log-log graph showing a rough estimate of the maximum sizes of organisms at different periods of life on Earth. (Data for the larger forms comes from Bonner, *Size and Cycle*, 1965; for data on the five smaller forms I am indebted to Shuhai Xiao)

WHAT CHANGES CAME FIRST: SIZE OR CONSTRUCTION NOVELTIES?

This brings us to the central argument of this book. If we look at the rising curve in figure 17 we see two things happening: there is an increase in maximum size over time, and there are great changes in the construction of the organisms involved.

This observation gives us an opportunity to ask which is the prime mover. What we find is that the universal selection for size is ever present, but in the case of size increase it is always limited by the internal construction of the organism. One could not build an organism even the size of a small mouse with a pile of amoebae. On the other hand—and we are thinking in terms of geological time involving great numbers of generations—there will be mutations that affect the internal structure. Those changes might be of such a nature that a larger size can be achieved and remain structurally sound. If that mass of amoebae trying to be a mouse developed some bone and muscle, it could not only exist but would have the basic building materials needed for the construction of an elephant. In other words, an increase in size requires adequate building materials, and until those are invented there is a strict limit to how big an organism can become through natural selection. This restriction means that size increase encourages and exploits innovations in structure, and those changes in structure do not cause but permit a greater increase in size.

To give another example of such innovations, the beginnings of vascular tissue in plants presumably served the function of supporting small stalks for more effective dispersal of spores or seeds and for that reason was selectively advantageous. Once a vascular system was established with a sound genetic basis, it could, through progressive modifications, ultimately produce giant trees. Therefore, it is not strictly true that size always precedes structure; clearly, the reverse is also true, and during the course of evolution there must have been a seesawing back and forth. Structure keeps pace with size, and size takes advantage and exploits every change in structure. When size changes lead the way, the structural changes may be relatively minor, often involving mere changes in proportions. However, when structural changes come first, they are often major changes that permit large increases in size, as can be seen when we follow the changes in maximum size in different major groups of animals and plants through the ages.

One thing is obvious in looking at the figure: the upward ride is in reality rather bumpy and not a smooth curve. This is so because the world does not consist of one species of different sizes, but many species with different body plans. Each of the separate groups of organisms will always find room at the top and therefore show a trend towards a greater maximum size, but each will be limited by its ancestral body plan. Think of the construction differences between trees and mammals or, on a less obvious level, the differences between birds and mammals. Selection for size will operate in each group, and for each efficiency will be maintained by natural selection. Note that the curve in figure 17 levels off as one approaches the world of today. Does this mean that we have reached some sort of upper size limit and that new or drastically modified body plans will be needed to evolve even larger animals and plants? (Before jumping to such a conclusion one must take into account that the graph is on a logarithmic scale.)

Size is a key object of selection, requiring changes in shape that will vary for different body plans. If there is a selection for a change in shape alone, then a change in size does not necessarily follow. For instance, an evolutionary increase in the thickness of the legs of a mammal would not require a larger body, while a larger body absolutely requires thick legs.

CELL SIZE

Despite the extremes one sees in the cell size of various small, unicellular organisms in figure 6, there is a remarkable constancy in the size of cells for most organisms. The biggest exception is bacteria; they indeed are much smaller than nucleated cells (eukaryotic cells), but then cyanobacteria (what used to be called blue-green algae) have cells comparable in size to eukaryotic ones. An exception for the largest cell in figure 6 is *Bursaria*, the huge ciliate. However, ciliates are special and in some sense can be considered equivalent to multicellular organisms. One can see the enormous "C" shaped nucleus in the figure, and this "macronucleus" has the genetic material duplicated many times. So it is functionally multinucleate, which has made it possible for ciliate protozoa to become large cells in this unique fashion.

Sometimes cells appear larger for other reasons. For instance, before development, bird eggs are enormous, but like most eggs they are loaded with yolk to feed the emerging embryo. Many plant cells seem exceptionally big because they have a huge vacuole containing water. In both cases the amount of cytoplasm is appropriately small for the size of the nucleus.

This brings us to an interesting point. The size of the nucleus usually reflects the size of a cell; there is a tendency to keep the nucleo-cytoplasmic ratio constant. For many years I had a gifted colleague, Gerhard Fankhauser, who did some fascinating work on newts.[13] By treating their eggs with temperature shocks at the beginning of their development, he was able to produce newts that had a different number of sets of chromosomes. Normally, most animals, including ourselves, have two sets, one from our mother and one from our father. We are diploids (2n), but Fankhauser could produce newts of different ploidies, from haploid (n) up to pentaploids (5n). For each there was a corresponding cell size (keeping the nucleo-cytoplasmic ratio constant), but the surprising thing was that the bodies of the grown-up larvae were all the same size, regardless of their cell size. This prompted him to look at the cells of these different individuals, and he discovered the remarkable fact that since the organs were all the same size, too, they were built with many small cells in diploids, compared to pentaploids where few, large cells were involved (fig. 18).

One day I suggested to him that it would be interesting to test newts of different ploidies in a learning maze since they would have the same size brains and therefore a different

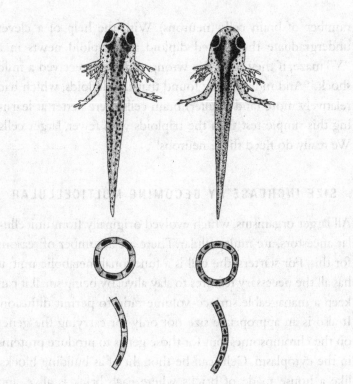

Figure 18. Two larvae of the salamander *Triturus viridescens*. The pentaploid larva with five sets of chromosomes in each cell (left) is the same size as the diploid larva with two sets (right). As can be seen from the circular cross section of a kidney tubule (middle) and a covering layer of cells over the lens (bottom), the tissues are also the same size despite the striking differences in cell size. (After G. Fankhauser; from Bonner, *The Evolution of Culture in Animals*, 1980; drawing by Margaret LaFarge)

number of brain cells (neurons). With the help of a clever undergraduate they tested diploid and triploid newts in a "Y" maze; if they took the wrong turn they received a mild shock.[14] And indeed, they found that the diploids, which had relatively more (and smaller) brain cells, were better at learning this simple test than the triploids with fewer, larger cells. We really do need those neurons!

SIZE INCREASE BY BECOMING MULTICELLULAR

All larger organisms, which evolved originally from unicellular ancestors, are multicellular. There are a number of reasons for this. For starters, the cell is a functional metabolic unit; it has all the necessary features to stay alive. By being small it can keep a manageable surface-volume ratio to permit diffusion. It also is an appropriate size not only for carrying the genes on the chromosomes, but for those genes to produce proteins in the cytoplasm. Cells can be thought of as building blocks, like a house made of bricks where each brick is alive, and together they make it so the individual can digest and process energy, move, orient itself in the environment, and in the case of animals, even behave and think.

I argued that there has been, during the course of evolution, a selection for larger and larger organisms because there is always room for bigger ones to escape the competition with the smaller ones. This is the case both within a species and among species. (In animals, size differences can also be driven by sexual selection and produce considerable size differences between the sexes, as is found, for instance, among some species of seals. To give an extreme example, elephant seal males are ten times the weight of females.)

The top of the scale is always an open ecological niche. This is the case right from the very beginning of multicellularity in early earth history: the easiest way to become bigger has always been to add more cells, and the size of those cells has always remained roughly the same from minute algae and invertebrates to giant animals and plants. The former may have as few as a dozen or a few hundred cells; there are considerably more than a trillion cells in our bodies, and since the number of cells is determined by the volume of an animal, imagine how many cells there must be in *Brachiosaurus* or in a blue whale.

It has been of great interest to me for some time that the invention of multicellularity did not just occur once but many times in the history of our earth. I was raised on the idea that it occurred once, or at the most twice, for the origin of animals and the origin of plants. My early student career in a laboratory was concerned exclusively with lower organisms, which forced me to realize the obvious: fungi, algae, slime molds, slime bacteria (myxobacteria), cyanobacteria, and a number of others could not possibly have come from one ancestor but from separate ones.

Furthermore, there are two different ways of becoming multicellular: the most common way would be for cells to divide and the daughter cells to stick together. Such a beginning would have been the norm in aquatic organisms and found in most of the groups mentioned above. However, there are a number of organisms that have another way. For them, feeding occurs as separate cells that then come together to form a multicellular organism. This is true for cellular slime molds, where the separate amoebae feed by engulfing bacteria; after they have cleaned an area, they aggregate to form a multicellular, unified, and coordinated individual. The equivalent method is found in other organisms: the myxobacteria, most fungi, and even in a unique multicellular ciliate. The interesting thing is that all these cases of multicellularity by aggregation are terrestrial, so there is a fundamental difference between them and the way aquatic forms became multicellular. This must have been the first conquest of land, long before it occurred in vascular plants, insects, and amphibians.

Each example has its own particular details of its construction. For terrestrial forms the cells differ; from rod-shaped bacteria, to amoebae, to fungal filaments, to ciliate protozoan cells. (In the case of fungi and true slime molds—myxomycetes—there are no cell membranes around each nucleus, as is the case in most other organisms. They are multinucleate, but as the great nineteenth-century botanist

Julius Sachs pointed out, each nucleus, and its surrounding cytoplasm, was a functional cell, an "energid" as he called it.) In aquatic forms, the cells may be bacterial, such as in the cyanobacteria, similar to the vast multitude of algae that also have stiff cell walls and are photosynthetic, or amoeboid cells with thin cell membranes, as in sponges and all the more complex animals. Then there are differences in how the cells stick together: in some cases their entire surface is sticky; in others there are specific sticky spots that affect the arrangement of the cells and therefore the architecture of the organism. One could go on with the details, but the message is that these differences are the basis of the great variety of living designs of multicellular organisms—large and small—that exist today.

It is important to ask what might be the advantage for these unicellular organisms to become multicellular, especially considering how successful they have been as single cells. It is true that it is an easy way to get larger, but what might be the immediate advantages for them? This is not quite like asking what might be the benefit to an elephant or a large tree; the reasons for the first steps towards multicellularity must be quite different. In a few cases we have some clues as to what some of them might be. Let me give three examples.

First, there are the myxobacteria, which are multicellular; what possible advantage might they have over their unicellular relatives? In some species they stick up into the air, making it possible to spread their cysts, which would be an advantage in dispersal. The more obvious advantage comes during their feeding stage. The individual feeding cells are present in multicellular swarms, and each bacterial rod secretes extra cellular digestive enzymes that break down complex food substances such as starches and proteins into their component sugars and amino acids in the same way our gut enzymes process the food we eat. They then can directly absorb those smaller energy-giving molecules. By having many such cells in a swarm, they can collectively produce enough of the enzymes to digest the particles around them; their large number means they can overpower potentially indigestible sources of food. This process has been called the wolf pack principle whereby many can capture the prey while lone cells will fail.

A second example among bacteria are the species that form clumps. These species live in places where there is no oxygen; oxygen will kill them. It is conjectured that if some oxygen should appear around a clump it will not get to the cells in the center, which will be protected and be able to survive.

A third example is illustrated by an interesting experiment on a unicellular alga that was grown in culture in the laboratory.[15] If one adds another species, a predator of the alga in the culture, then all the small cells will be eaten. However, occasionally a mutation occurs in the vulnerable alga that produces a sticky substance on its surface, with the result that the cells form clumps that are too big to be engulfed by the predator, allowing them to survive.

There are many other ways in which togetherness produces obvious advantages: the important point is that reasons for becoming multicellular are easy to conceive, and all depend on the advantage of a collective behavior—a size increase.

BACK TO LEMUEL GULLIVER

Because of our modern understanding of the microscopic, cellular nature of animals and plants, we see that there are some biological problems with Jonathan Swift's delightful size worlds. The cells of the Lilliputians, of Gulliver, and of the Brobdingnagians would all have to be the same size; therefore, as one went up the size scale, the characters would have been made up of more and more cells. This would mean, among other things, that the Brobdingnagians would be brainier than the lesser mortals—following the arguments of Bernard Rensch discussed earlier—but such a progression of intelligence is not clear from the book. The eyes of the Lilliputians would have the same size cells as their larger brethren, and therefore their eyes would be made up of fewer cells. Unfortunately this would not mean that they could see smaller things, as though their vision was through a microscope: they would see small objects with no more acuity than Gulliver or the Brobdingnagians. If anything, they would have a smaller field of vision that would put them at a relative disadvantage. Their plight is clearly the reverse of Gerhard Fankhauser's newts of different chromosome numbers: in Jonathan Swift's world the cells stay the same size but the bodies vary from huge to small, and in the newts body size is constant while the cell size varies.

As we have seen, there are many important consequences to being small, and the cells found in all organisms fall in that lower size realm. However, by sticking together they have been able to produce large multicellular animals and plants, and in doing so they have crossed from one size world to another.

CHAPTER 5

SIZE AND THE DIVISION OF LABOR

For many years I have been struggling to understand the relation between size and the division of labor in multicellular organisms. One wants to know why it is generally true that the bigger the organism, the greater the division of labor. It is a principle that to some degree we all accept as given; let us examine it closely.

The division of labor, which is a reflection of biological complexity, is a venerable subject. The history of the idea has a few years ago been admirably discussed by Camille Limoges.[16] He points out that the pioneer for considering the matter within organisms goes back to Henri Milne-Edwards, who as early as 1827 said that animals had a "division of physiological labor." Limoges goes on to say that Darwin was favorably impressed by the writings of Milne-Edwards, who in turn influenced the pioneer in sociology, Émile Durkheim. The conclusion of Limoges is that there has always been a close association between the concept of division of labor as it applies to individual organisms and as it applies to societies (a matter to which I will return). However, none of these earlier authors were particularly interested in size as a significant correlate. This tradition continues to this day, and the matter of size is not central to our current thinking. What I wish to show here is that significant increases or decreases in size during the course of evolution are accompanied by increases or decreases in the division of labor.

We now come to another of our basic size rules. The rules discussed so far are the relation between (1) size and strength, along with (2) size and surface areas. There is a similar relation between size and complexity. Like the others, it also can be expressed as an allometric relation:

$$(3) \text{ Complexity} \propto \text{Weight}^a.$$

As we proceed from the strength-size rule, and particularly to the surface-size rule, there is a direct trend towards the complexity-size rule. As I pointed out before, in a larger aerobic animal it is imperative to get oxygen to the internal tissues. To do this it is necessary to produce a circulatory system with a pumping heart, innumerable capillaries, and lungs or gills with huge surfaces to trap the oxygen that is then carried by the red blood cells to the tissues. Plainly, as size increases the demands of the surface-size rule has dictated an enormous increase in complexity. The three rules are indeed connected to one another and they are all directly related to changes in size.

A convenient way to measure biological complexity is to think of it in terms of the number of cell types. A mammal will have a large number of distinct cell types, such as muscle cells, liver cells, brain cells, blood cells, and so forth. A parallel kind of list can be made for all other organisms, including plants. The number of cell types can be considered a rough reflection of the division of labor, or complexity; it is a useful indicator.

For each rule, an increase in size requires an increase, be it in strength, or in surface, or in the division of labor. On the other hand, a decrease in size generally does not necessarily require a decrease in these three properties; rather, they are simply permitted, for the smaller organism is likely to be able to manage even though it might have a surfeit of strength, or surface, or division of labor.

Estimates of the number of cell types for different size organisms have been gathered from the literature by Graham Bell and Arne Mooers.[17] Their results are an excellent way of illustrating the size-complexity rule as can be seen when they are brought together in a graph (fig. 19). The larger the organism, the greater the number of cell types.

Note that despite the clear trend there is a wide scatter of the points in the figure. The variation might have a number of causes. It is very difficult to make accurate measurements of the number of cell types; all these estimates are rough. The data come from different authors who undoubtedly have differences of opinion as to what constitutes a discrete cell type.

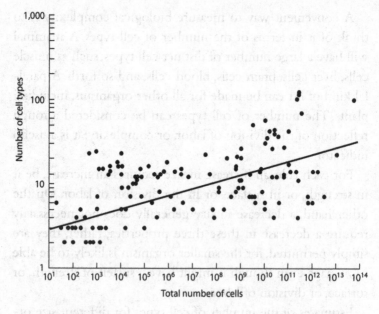

Figure 19. A log-log graph of the number of cell types plotted against the total number of cells for a vast array of organisms from very small colonial forms to the largest animals and plants. (Redrawn from G. Bell and A. Mooers, *Biol. J. Linn. Soc.* 60 [1997]: 345–363)

It should be noted here that the size-complexity rule also applies during development. Both plants and animals normally begin at a single-cell stage and become increasingly multicellular with time. During this process there is a proliferation of cell types, which ultimately become adults that reach the number shown on the curve in figure 19. Since during development the organism becomes progressively larger, it would be very interesting to know how many cell types there are at each developmental size. Unfortunately, I have been unable to find any such measurements in the literature, but it is obvious that the size-complexity relation must hold. However, there is one point to remember: many of the cell types will not be functional when they first appear. For instance, all the cell types associated with digestion in the gut in an animal are passive until the first meal. Before that they are merely preparing for the passage of food. So the size-complexity rule is somewhat different during the course of development, yet the rule holds.

THE ORIGIN OF DIVISION OF LABOR

My plan here is to examine the size-complexity rule across a huge spectrum of sizes, from small and lowly primitive organisms to huge societies. The former are of particular interest because they give some insight into how the division of labor might have arisen in early evolution. There is no better place to examine the matter than at the origin of differentiation, where one finds closely related species with either one cell type or two. An example is found among the green alga *Volvox* and its relatives.

Volvocine Algae

Species of *Volvox* are common in freshwater ponds, first described by van Leeuwenhoek in the seventeenth century using his simple microscope. It consists of a sphere of a single layer of some thousands of cells at the surface of a sphere (rather like a minute tennis ball with its hollow center) about

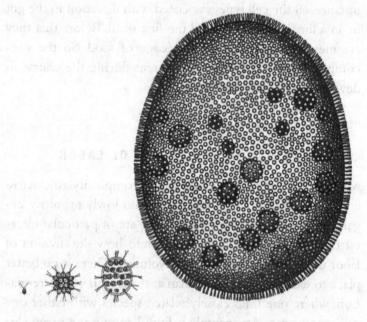

Figure 20. The ranges of sizes in the volvocine algae. From left to right, the ancestral single cell *Chlamydomonas*, the 16-cell *Gonium*, *Eudorina*, and *Volvox*. (Selected from W. H. Brown, *The Plant Kingdom*, 1935)

the size of a small pinhead (fig. 20). Most of the cells have two flagella that stick outward and beat in coordinated waves so that the whole ball rotates like a globe that can spin around its axis. What concerns us here is that in *Volvox* there is a division of labor: most of the cells are the ones just mentioned that move the colony and are capable of photosynthesis, but there are a few cells that are reproductive and have no flagella—they develop into daughter colonies.

A number of other species of volvocine algae are smaller and are made up of fewer cells (fig. 20). The whole group has been studied in exemplary detail by David Kirk, and among many of the things he and his co-workers discovered using DNA analysis, the multicellular forms arose from a single-cell ancestor known as *Chlamydomonas*, as had been suspected before based on morphological grounds.[18] Another interesting bit of information has come out of Kirk's DNA analysis of the ancestral tree of the volvocine algae. The old, conventional idea that they evolved by a simple, linear progression from small to large has been shown to be false. There is indeed an obvious span of sizes from the unicellular *Chlamydomonas* to the largest species of *Volvox*, which may consist of 50,000 or more cells (fig. 20). However, the history of the group is more complicated than was previously believed. *Volvox* itself has arisen independently more than once during the course of evolution. Even more relevant to this discussion, it appears that certain species of *Volvox* are ancestral to smaller forms, such as *Pleodorina*, as well as the reverse. So there apparently have been evolutionary decreases as well as increases in size during volvocine evolution.

Here I want to focus on the two cell types that are present in the asexual life cycle of *Volvox*. The biflagellate, non-reproductive (somatic) cells are responsible for the movement of the colony, and the reproductive cells (gonidia) produce daughter colonies. The former are terminally differentiated and are incapable of dividing, while the latter become cell-division specialists and are incapable of locomotion.

There is another important bit of information concerning this dichotomy. We know that very few genes are involved in this transition from one to two cell types. The initial discovery was the mutation of one gene that would transform a normal *Volvox* so that all the cells that would be somatic became reproductive and divided very rapidly, producing thousands of minicolonies in one large individual, similar to what occurs in small volvocine species. Subsequently, it was discovered that a few other genes were involved, but clearly the invention of this division of labor appears to involve only a small number of genetic steps.

This two-cell-type state is correlated with size. At one end of the size range, *Volvox*, which contains thousands of cells, always has both cell types. At the other end of the spectrum, *Gonium*, which never has more than 16 cells, has no division of labor: all *Gonium* cells are initially biflagellate and involved in locomotion, but later all of them shed their flagella and undergo reproduction by successive divisions to produce daughter colonies.

There is a particularly interesting feature concerning the invention of the division of labor among the volvocines. Some species can exist in different sizes, and only the larger forms will have more of the two cell types. They apparently can asses their size: the larger they are, the greater the division of labor. This ability is known as *quorum sensing*.

Eudorina can have either 16 or 32 cells, depending on environmental conditions. In the 16-cell colonies, like *Gonium*, all of the cells participate first in motility and then subsequently in reproduction. However, in 32-cell colonies, the four cells at the anterior end of the colony often remain terminally differentiated nonreproductive cells that continue beating their flagella, while the other 28 cells reproduce. In another genus, *Pleodorina*, which always has a division of labor, the ratio of nonreproductive cells to reproductive cells increases as the total cell number is increased. In colonies with 32 cells, 25 percent of the cells are sterile nonreproductive cells, while 50 percent of the cells are so in colonies of 128 cells. Clearly, within this critical intermediate size range the proportions of the two cell types varies depending on the size of the colony.

A central question is what selective advantage might volvocine algae have gained by producing terminally differentiated somatic cells that are incapable of reproduction. It might be simply that this dichotomy enables them to reproduce while simultaneously swimming and gaining energy through photosynthesis. If it is a small species that has only one cell type, in its nonmotile reproductive phase it will sink to the dark bottom of the pond where photosynthesis would be impossible. In *Eudorina* and *Pleodorina* we see an intermediate situation: in the larger forms there is an increase in ability to carry on photosynthesis while reproducing. This is the basis for suggesting a quorum-sensing mechanism where a group of cells take into account their number so that they behave differently when they are large as compared to small. It means that in these instances there must be a genetic mechanism for setting apart the two cell lineages, a system that is not rigid but plastic and capable of responding to size. It is puzzling to imagine in this case what the selective advantage of quorum sensing might be; perhaps there is none and it is simply a by-product of the genetic activity.

Slime Molds

The cellular slime molds provide another interesting example of the borderline between one cell type and two. They have an unconventional life cycle in that the separate amoebae eat bacteria in the soil, and upon starvation they aggregate to form a multicellular individual that ultimately produces a small fruiting body: a mass of encapsulated spores held up by a delicate stalk. Unlike most other organisms, they feed first and then become multicellular. Most species of slime molds have two cell types: spores and stalk cells. The exceptions are the species of *Acytostelium* that make only spores; their stalk consists of a delicate strand of cellulose extruded by all the cells before they become encapsulated into spores (fig. 21). From some work directed by S. Baldauf and P. Schaap using DNA methods to build a slime mold ancestral tree, we now know that *Dictyostelium discoideum* (including some of the other common *Dictyostelium* species) and *Acytostelium* are far removed from one another and only distantly related.[19] So, unlike what we saw in the volvocine algae, the two genera of slime molds did not go back and forth in size during their evolution.

Nevertheless, the different species of *Dictyostelium* that have two cell types are in general larger than *Acytostelium* that has no stalk cells and therefore only one cell type. There is a separation between the smaller species with one cell type and the larger ones with two. In these organisms, size is not determined by multicellular growth, as it is in most organisms including *Volvox*, but by the number of amoebae that enter an aggregate. Therefore, this size-complexity relation holds regardless of how size increase occurs. Despite this difference, both the volvocine algae and the cellular slime molds are similar in having a size threshold for the number of cell types.

Quorum sensing is also found in one species of cellular slime mold; the case of *Dictyostelium lacteum* is particularly interesting. When very small fruiting bodies arise, a partially

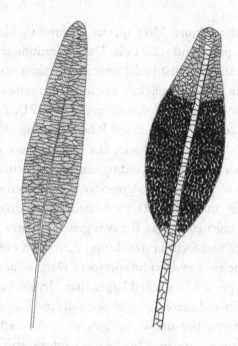

Figure 21. A diagrammatic comparison of a rising fruiting body of *Acytostelium* with only one cell type (left) and *Dictyostelium* with two—the dark spores and the stalk and prestalk cells (right). A particularly small *Dictyostelium* is drawn so that the cell structure of the two can be compared. (From Bonner, *The Evolution of Complexity*, 1988)

acellular stalk is produced;[20] it is as though there is a threshold concentration of a key substance or substances that stimulate a cellular stalk, but if there are very few cells, the concentration is insufficient to be wholly effective in producing a stalk with cells (fig. 22).

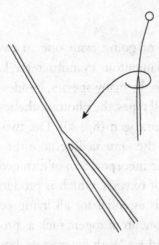

Figure 22. A very small fruiting body of *Dictyostelium lacteum* showing an acellular tip. (From Bonner and Dodd, *Biol. Bull.* 122 1[1962]: 13–24)

Why is there quorum sensing involving the division of labor? It might be adaptive, for it reduces the cost of construction in those forms that are of intermediate size. In the case of *D. lacteum*, for instance, if a cellular stalk is not needed for support, then, by making part of the stalk acellular, a few extra spores are produced. In larger species, very small fruiting bodies can be produced in which the stalk is cellular; the smallest one known is a fruiting body of one species (*P. pallidum*), made up of three stalk cells and four spores. In other words, quorum sensing is not universal and not found in larger species but only in one species near the borderline between large and small.

Cyanobacteria

A final example of going from one to two cell types may be seen in the filamentous cyanobacteria, known to be very ancient organisms. In many species, besides spores they have two vegetative cell types: the photosynthetic cells and the heterocysts that fix nitrogen (fig. 23). The two activities cannot be carried on in the same cell because the enzymes that are responsible for the incorporation of nitrogen cannot function in the presence of oxygen, which is produced by photosynthesis. Nitrogen is essential for all living cells for many key substances that contain nitrogen, such as proteins and nucleic acids and many others. With heterocysts, both of the essential biochemical activities can occur at the same time.

Not all species have heterocysts: some have a temporal separation of the two activities, a temporal division of labor. All the cells photosynthesize during the day and fix nitrogen at night.

In cyanobacteria, the ratio of vegetative cells to heterocysts remains constant regardless of the length of the filaments. The important size constant is the physiologically functional units;

Figure 23. A cyanobacterium filament showing evenly spaced heterocysts. (From G. M. Smith, *The Freshwater Algae of the United* States, 1933)

they are modules. This is a form of quorum sensing, for a heterocyst senses the number of neighboring vegetative cells, and a proportion is retained in each module. In this case, the selective advantage is obvious: to produce a balance between photosynthesis and nitrogen fixation.

Quorum Sensing

Quorum sensing is a newly recognized phenomenon that has become an exciting field of research in microbiology. The idea is that when cells come together, a critical number is required to be able to perform activities that would not be possible if there were fewer cells. I have already described a somewhat similar phenomenon in the wolf-pack feeding of myxobacteria, where feeding can be effectively achieved only if large swarms of cells exist; only then can they produce enough extracellular enzyme to effectively digest external food. Bonnie Bassler made the key discovery that luminescent bacteria are only capable of giving off light if they are present in a critical mass—when that number is attained, the light goes on.[21] This has turned out to be of great importance in medicine because the same situation is found among pathogenic bacteria that can only produce their undesirable effects when they are present in sufficient numbers: a quorum is needed before they can do their damage.

Both the volvocine algae and the cellular slime molds provide good examples of size quorum sensing. In each case they occur in species that are intermediate in the size range, and the individuals of those species can vary in size. If the individuals are large they will have two cell types, if small only one. They can sense their size and provide and respond accordingly when they are above or below a certain threshold.

In all the cases we have discussed with one or two cell types, including those cases of quorum sensing, it is evident that size is tightly correlated with—and very likely the direct cause of—the number of cell types, and not the reverse. Changing the number of cell types does not dictate the size of the colony or group—it is the size that dictates the number of cell types.

SIZE DECREASE IN LARGER ORGANISMS

The lesson to be gained from this kind of quorum sensing is that there can be a tight control between size and the number of cell types. If we now turn to larger organisms with many cell types, we find a somewhat more general correlation, but nevertheless the size-complexity rule holds. We have already seen that when there has been a size increase during the course of evolution, it has often been accompanied by an increase in the number of cell types, but even more revealing in providing evidence that a change in size can result in a change in the division of labor are those cases where there has been a decrease in size and a corresponding decrease in the number of cell types. Here an example will be given for animals and another for plants.

Rotifers

Rotifers are very small animals that live in ponds and lakes and other wet places (see the rotifer in fig. 6). They feed on small unicellular forms, especially bacteria. The majority of rotifers have a conventional gut tube, while others have a central part of their body into which food vacuoles are formed at one end and excreted at the other, very similar to the digestive system found in ciliate protozoa (fig. 24). The no-lumen rotifers lack a cellular stomach and on this basis are considered to have fewer cell types.

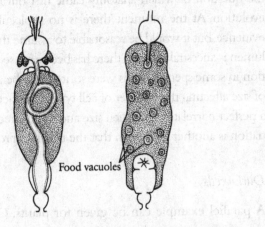

Food vacuoles

Figure 24. A diagrammatic view of the gut of a species of rotifer, one with a lumen (left), and one without a lumen (right). (From T. A. McMahon and J. T. Bonner, *On Size and Life*, 1983)

In the past it has been my assumption that their ciliate-like digestive system is related to small size. This was just asserted; is it possible to find any evidence for the assertion? Fortunately, Josef Donner in his exhaustive monograph of the Bdelloid rotifers gives the length of large numbers of species,[22] and I have extracted this information for forty-eight species without lumen and seventy-four with. It is clear that the mean lengths of the former are significantly smaller than the latter, and the assertion is justified.

However, the overlap is considerable: some species without a lumen are larger than the smaller ones that have one. This raises the question of which anatomy came first during the course of evolution. At the moment there is no molecular ancestral tree evidence, but it would be reasonable to assume that the gut with a lumen is ancestral and that there has been a subsequent size reduction in some species. If this were so, it would be another example of size affecting the number of cell types. The fact that there is not a perfect correlation between size and the degree of cell differentiation is another indication that the rule is approximate.

Duckweeds

A parallel example can be given for plants. Ordinary angiosperms are estimated to have up to forty or more cell types while the two small duckweeds have far fewer: *Lemna* has eighteen and the minute *Wolfia* about eight (fig. 25). There is little doubt that these smaller forms are descendants of larger

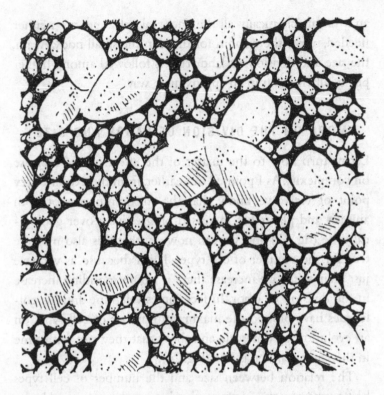

Figure 25. Two species of duckweed that are the smallest flowering plants. Floating on the surface of water, the minute *Wolffia* (about 1 mm long) surrounds the larger *Lemna*. (Drawing by Hannah Bonner)

ancestors and that there has been a reduction in the number of cell types accompanying the size decrease. The key question is, why are there fewer cell types? The most obvious explanation is that there has been a selection for size decrease, and that mechanically all the ancestral cell types are neither needed, nor is there room for them in the small body. Again, the size and division-of-labor rule is followed among higher plants, albeit again in an approximate way.

MORE ON THE DIVISION OF LABOR AND SIZE

Let us turn now to the matter of the effect of size increase on complexity. As I pointed out earlier, from the evolutionary point of view we know that the maximum size of both animals and plants has increased progressively over geological time (fig. 17) and we can now say that this also must be true for the number of cell types. This is because, as we have just seen, with size increase there is a corresponding increase in the division of labor. In fact, James Valentine and his colleagues have plotted the maximum number of cell types on an evolutionary timescale and show that they rise with time in the same manner as size (Fig. 26).[23]

The relation between size and the number of cell types being approximate, one may ask what sets the upper and lower limits of the size range for organisms with a particular number of cell types. If, in the selection for size changes, an animal or a plant becomes too big or too small to be physiologically efficient and be able to compete effectively, it will become extinct. The range of size among mammals is extensive—from mice and shrews to elephants and whales—yet each in its own

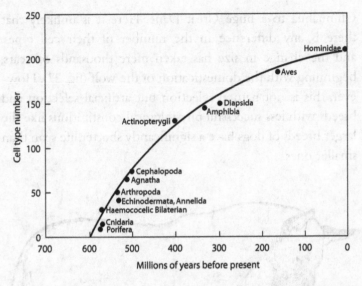

Figure 26. Estimated numbers of cell types of early members of various animal groups. Only members of the groups that are believed to have been rather near the upper limit of cell type numbers when they originated are included. (From Valentine, et al., *Paleobiology* 20 [1994]: 131–142)

environment can successfully exist, and they do not differ to any great extent in their number of cell types. This great span of sizes of similarly built and physiologically efficient animals has taken millions of years to achieve—each step requires adjustments for stability.

A large size span can, however, be achieved very quickly through artificial selection. Consider dogs, from the miniature

Chihuahua to a huge Great Dane. Here it is unlikely that there is any difference in the number of their cell types, and the change in size has taken mere thousands of years, beginning with the domestication of the wolf (fig. 27). However, this is not natural selection but artificial selection and breeds with less successful physiological constitutions like the larger breeds of dogs have a significantly shorter life span than smaller ones.

Figure 27. Showing the span of the sizes of dogs, the result of artificial selection for size from the ancestral wolf. A large Great Dane alongside a small Chihuahua. (Drawing by Hannah Bonner)

Energy and Size

This leads to an important point that has always puzzled me. Why do mammals with the same number of cell types have a huge size range? The answer comes from an important insight of Megan McCarthy and Brian Enquist, who show that it has to do with their levels of metabolism.[24] Size is not the only factor to influence the number of cell types, but metabolic energy is an additional key factor that correlates with cell type numbers. The metabolic rate of a mouse is much greater than that of an elephant, and for this reason it also supports a large number of cell types. A related question is, why do animals have more cell types for a given size than plants? The reason is the same: plants have a much lower metabolic rate than animals. In the next chapter I will return to this relation of metabolism to size.

Societies

The whole idea that there is a relation between size and the division of labor was first realized in human societies. It is an ancient principle that we associate with the eighteenth-century economist Adam Smith. The larger a nation or a community, the greater the division of labor, often expressed in terms of an increase in specialized trades or occupations. In a small village, individuals have to be jacks-of-all-trades, while in larger communities there is a cobbler, a butcher, a smith, a tailor, and so forth. In even larger cities or towns, there is a further division of labor where we find lawyers, bankers, policemen, doctors, and many other specialized occupations. Even institutions as they increase in size will divide the labor: large businesses will have a whole hierarchy of managers; large universities will have layers of administrators and teaching faculty; an army has a multitude of officers and enlisted men of different ranks, and so forth. The rule clearly holds for human societies, and there are a few examples where measurements have been made (fig. 28a, b).

The rule also applies to insect societies, which are quite remarkable for their size and diversity. Some ant colonies will have one queen with millions of workers attending to all the activities and functions of the nest, such as feeding and caring for the young, foraging for food, and protecting the colony. At the other end of the spectrum there are certain primitive wasps that have very small colonies, consisting sometimes of less than a dozen individuals.

Figure 28 (facing page). Log-log graphs showing the relation of size to the division of labor for societies. (a) The number of occupations found in each of the states in India (graph of N.V. Joshi in Bonner, *Current Science* 64 [1993]: 459–466). (b) The number of organizational traits (specific crafts and occupations) plotted against the size of single-community societies. (Five of the latter are labeled.) (From Carniero, *Southwest Journal of Anthropology* 23 [1967]: 234–243)

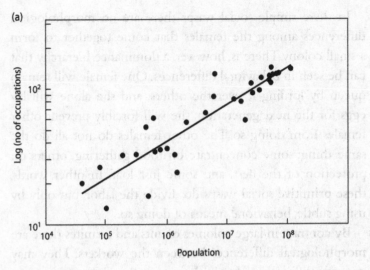

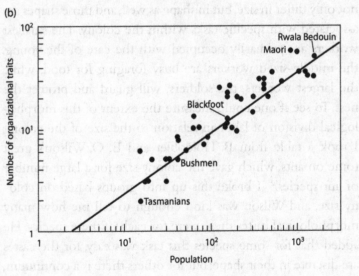

In these simple social wasps there are no morphological differences among the females that come together to form a small colony. There is, however, a dominance hierarchy that can be seen in behavioral differences. One female will remain queen by lording it over the others, and she alone will lay eggs for the next generation: she will forcibly prevent other females from doing so. The other females do not all do the same thing: some concentrate on food gathering, others on protection of the nest, and some just loaf. In other words, these primitive social wasps do divide the labor, but only by using subtle, behavioral means of doing so.

By contrast, in large colonies of ants and termites there are morphological differences between the workers. They may not only differ in size, but in shape as well, and those shapes are associated with specific tasks within the colony. The smallest workers are primarily occupied with the care of the young, the middle-sized workers are busy foraging for food, while the largest workers, the soldiers, will guard and protect the nest. To see if one could measure the extent of this morphological division of labor in relation to the size of the colony, I took a table from B. Höldobler and E. O. Wilson's great tome on ants, which gave the colony size for a large number of ant species.[25] I broke this up into groups based on colony size, and Wilson was kind enough to tell me how many morphological castes there were for each of these species. He added that for some species this task was easy, for the castes are discrete in their shape; but for others there is a continuum,

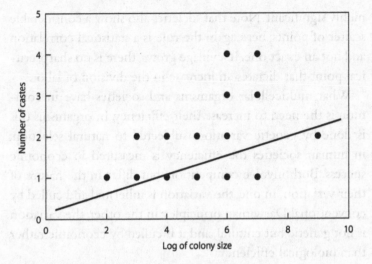

Figure 29. The number of ant castes plotted against the size of the colony. The line is a regression showing a significant trend ($r^2 = .25$). (From Bonner, *Current Science* 64 [1993]: 459–466)

and therefore assigning a number for the different castes is difficult and subjective. I was fortunate to be able to rely on his great experience in observing ants and was able to plot colony size versus caste number to show that, as expected, there is a clear and significant increase in the number of castes with an increase in the size of the colony, despite a considerable scatter of the points (fig. 29).

The results for societies can be compared with similar plots for multicellular organisms (see fig. 19). In all cases the trend is highly significant. Note that societies also show a considerable scatter of points; here again the rule is a statistical correlation and not an exact one. If a village grows, there is no sharp, critical point that dictates an increase in the division of labor.

What multicellular organisms and societies have in common is the need to increase their efficiency. In organisms this is done by genetic variations subjected to natural selection; in human societies the efficiency is measured in economic success. Both involve competition, but differ in the nature of their variation. In one, the variation is inherited and culled by conventional Darwinian principles; in the other, the variation is not genetic but cultural, and it is culled by economic rather than biological efficiency.

Thus far we have examined three size rules: strength and diffusion keep pace with size increase, as does the division of labor. The latter is not only true for the number of cell types within an organism, but for the number of specialized activities in societies. The three rules are closely interrelated and show the ubiquitous importance of the role of size.

In looking at the relation between size and the division of labor, the important question is the effect they have upon each other. Natural selection will act on both size changes and the internal physiological efficiency. If there is selection for a significant increase in size, selection will in turn change the internal structure to maintain physiological efficiency, and this effect may involve an increase in the division of labor.

If there is a selection for a size decrease, again the efficiency must be maintained. Usually this process does not require a decrease in the division of labor, although, as has been shown, it can occur. Going in either direction there are presumably thresholds beyond which efficiency (involving the appropriate division of labor) must change to keep pace with size.

This means that within these limits there can be changes in size without changes in complexity, and conversely changes in complexity without changes in size. Because of the latitude of change within the limits for both size and complexity, one can expect—and finds, as we have seen—large variation in the size-complexity rule.

If we now ask why human societies seem to follow the same size-division of labor principle as cells, the answer is somewhat different from that of cells. There are none of the obvious physical constraints we saw for cells; but even more important, there is a totally new element, namely behavior. Individual organisms devoid of a brain can only pass information through their genes, but in the evolution of animals there arose an entirely new method of passing information, and that is by behavioral signals between one individual and another. These signals have been called "memes" by Richard Dawkins.[26] Genes can be passed on only through egg and sperm (or asexual bodies such as spores) from one generation to the next, and therefore many generations are required over a long period of time for a particular gene to spread through a population. Memes, on the other hand, can pass from one individual to another in an instant of time, and as a result can spread very rapidly through a population.

There is, of course, much more to behavior than the passing of information between individuals; for instance, we can solve problems. Animals can also solve problems, but of all animals we are undoubtedly the most skilled in doing so. It is this problem-solving ability that is the key to how the labor is divided in societies.

A good example may be found in ant societies. Their division of labor is ultimately genetically based—including their ability to respond to nutritional and environmental conditions—and therefore subject to natural selection. This is certainly true, but insects have some flexible behavior as well, and perhaps even a modest ability to solve problems. Those who work with social insects can give many examples; here is one that makes my point. If, in an ant colony where different castes perform different labors, one removes one of the castes, then some of the workers from other castes will take over the tasks that they never performed previously. In other words, they show a behavioral flexibility; in a way, this could be considered an example of minor problem solving. By being flexible they see to it that all functions, all the labors, required for the welfare of the whole colony are carried out, despite the loss of one of the castes. From this one might conclude that in social insects there can be a mixture of behavior in a background of a strong genetic basis for the division of labor.

For human beings, the genetic component takes a back seat. No doubt the structure and the abilities of our brains have a genetic basis, but what we do with our brains in terms of thoughts, including problem solving, is way beyond our genes. Our skills may have a genetic basis, but what we do with those skills opens up a new world.

Today the whole subject of division of labor has taken on a new cast for human societies. With the tremendous increase in the size of our population, standard ways of increasing that labor, such as we have been discussing, have become increasingly complicated. This is true in all ways: in commerce and in economics in general, and in government. We are in an era of world trade, of globalization, of amalgamation of businesses into mega corporations; size alone has created vast complexity, and new ways of remaining efficient are being created. This has been helped, even fueled, by the extraordinary increase in the efficiency and rapidity of our methods of communication, beginning with the telephone, the radio, television, fax, and then the extraordinary burst created by computers with all the lightning powers of e-mail and the internet. World population and these technological advances have brought us beyond the straightforward size/division-of-labor relation I have been describing. Now we are in the next phase of unimaginable complexity. It is as though we have gone from the brain of an insect to that of a human, where not only has there been an enormous increase in the number of cell types, but the sheer number of neurons and their interconnections have risen astronomically. Just as our brains are capable of doing things that an insect cannot do, and does everything in a different way, so our modern society deals with all matters that preserve efficiency in a way that is different from what we did before.

From the many examples of different magnitudes that we have examined—from two cell types to huge societies—we see that the change in size comes first, followed by the degree of the division of labor. This process is especially clear-cut when there is a size increase that directly leads to a greater division of labor. But we also saw that a decrease in size can be followed by a corresponding reduction in the number of cell types, as in rotifers and duckweed. The most compelling evidence of size being the arbiter of complexity is from the quorum sensing in volvocine algae and in cellular slime molds, where only two cell types are involved. There is either one cell type or two types, depending on a small shift in size.

THE SIZE-ABUNDANCE RULE

We now go to a big canvas: to huge ecological communities. There we have organisms of all sizes living together and interacting with one another either in competition or in cooperation. In many ways it is similar to a community of cells that make up a multicellular organism: the component parts in each case are continually communicating and inter- acting with one another so that in both cases there is a collec- tive wholeness. In one case it is within an individual multicel- lular organism, and in the other it is between the individual organisms in a patch of nature. One major difference is of size: individual organisms get no bigger than whales and giant sequoias, while ecological communities can spread over huge geographic areas.

It is obvious that small organisms will be more abundant than large ones—we are surrounded by evidence of this fact. The number of bacteria per square mile in the African veldt is unimaginally larger than the number of elephants.

Beginning with plants, it is well known that trees in a growing forest first are a dense thicket of saplings, but as they grow they do not all survive; there is a "die-off" or "self- thinning" of the less fortunate individuals, so that a forest or patch begins with many small young trees, and ends up with a few large ones. The same is true of crop plants if sown too close together. Clearly, as the plants grow they need greater resources, such as leaf space to catch the sun for photosynthe- sis and root space to gather water and minerals in the soil; they compete with one another, and as a result there are winners and losers.

The same rule applies to animals; the main difference is that animals are mobile. The spacing of animals is also governed by energy considerations and the availability of resources in the form of food. Since animals do not use the sun's energy for nutrition but eat vegetable matter or meat, in general the consumer is larger than the consumed. Many of the larger animals, such as giraffes or moose, are plant eaters, and they certainly are larger than the salads they consume. This is true at the other end of the size scale, too: minute rotifers eat even smaller bacteria and unicellular forms. There are exceptions to this very general rule. For instance, large plants can also be consumed by small insects. And parasites have fiendish ways of attacking hosts many times their size, either by devouring them from within, such as parasitic wasps infesting huge caterpillars, or living within a large animal as a nonlethal partner, as for example our intestinal bacteria that help to process our food.

Earlier I laid down five size-related rules, and the relation between the abundance of organisms and their size was one of them. The rule can be expressed as

$$(4) \text{ abundance} \propto \text{Weight}^{-b},$$

which simply says the greater the size of an organism, the fewer individuals there are in a geographic area. (Note that contrary to the other rules, the slope of the line goes in the opposite direction.) This is an old subject that has been measured and treated in detail. Following the equation above, one

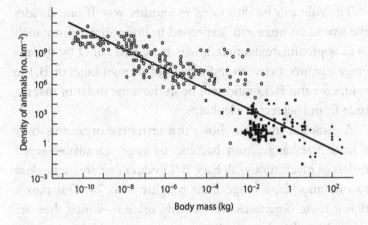

Figure 30. A log-log plot of the size of animals and their abundance in nature. Solid dots are mammals; squares are birds; hollow circles are invertebrates; x's are reptiles and amphibians. (From R. H. Peters, *The Ecological Implications of Size*, 1983)

can plot the logarithm of the density against the logarithm of the size of the organism. R. H. Peters has collected a large amount of information on animals from the literature and put it in one graph (fig. 30).[27] A straight line can be drawn through all the points, although as Peters discusses there is a considerable scatter of the points, making it difficult to be certain of any general mathematical principle. The important message for us is that the size-abundance rule holds for a tremendously wide range of organisms.

The rule can be illustrated in another way. If one divides the size of an organism (expressed in linear dimensions, such as its approximate diameter), which we will call D, by the average distance between individuals of its own kind, or B, the values for this B/D ratio will be in the same order of magnitude from bacteria to elephants.

A glance at the table shows that terrestrial organisms, over a huge size range from bacteria to large mammals—seven orders of magnitude—all have B/D values that are quite close to one another, a range from roughly 7 to 70. This means that if these organisms were evenly spaced—which they are not—then the distance between them would be more or less the same number of body diameters apart, no matter their size.

The same is undoubtedly true for plants, as one would expect from their "die-off." To give an example, my colleague Henry Horn has measured the trunk diameters and the distance between trees in five different woodland areas, and the B/D values are all within the same range as those of the animals.

Measuring the ratio of the distance between organisms and their linear size and finding that it is roughly constant over a great range of sizes is no more than another way of illustrating the size-abundance rule. Note that since both measurements are distance, that is, linear, the B/D ratio is a dimensionless number.

ANOTHER WAY OF VISUALIZING ANIMAL ABUNDANCE

By dividing the distance between organisms (B) and their diameter (D) it can be seen that over a size span of 7 orders of magnitude the ratio is moderately constant.

Organism	Size (in diameter)		Distance between	B/D	Source
	Range	Mean			
Bacteria	1–10μ	3μ	44μ	15	Waksman (1932)
Amoebae Arthropods	10–100μ	40μ	390μ	10	Singh (in Waksman, 1932)
Nematodes	0.1–1mm	0.5mm	5.8mm	12	Park et al. (1939)
Arthropods Other invertebrates	1–10mm	5mm	35mm	7	Park et al. (1939)
Invertebrates Shrews Small birds	1–10cm	5cm	50cm	10	Est. from Peters (1983)
Birds	10–100cm	10cm	300cm	30	Oelke (1966)
Large mammals	1–10M	1.5M	100M	67	Est. from Peters (1983)

CHAPTER 6

SIZE AND TIME

At the time the new millennium was about to come upon us, I was asked by the Buddhist magazine *Tricycle* to write a few paragraphs on what the year 2000 might mean to my slime molds.[28] This seemed to me like a fun idea and this is what I wrote:

TIME FROM THE POINT OF VIEW OF A SLIME MOLD

Time and life are intertwined in so many different ways, something all biologists are acutely aware of. Consider a few extremes: a single cell bacterium may have its entire life cycle in half an hour, but a generation for an elephant takes 12 years and a giant sequoia takes 60 years. One reason I work with slime molds, which are soil amoebae that start off as single cells, and then come together to form a multicellular organism, is that their generations are short, so that if I start an experiment on Monday, I will know the result by Wednesday or Thursday. This kind of biological time—life cycle time—is at the middle of the time scale of living phenomena.

At the faster end of that scale is physiological time: how many beats does a heart have in a minute, or how long does it take to swerve the car in order to avoid a squirrel on the road. As with life cycles, these rapid living activities are greatly affected by size, so a huge elephant will have about 25 heartbeats per minute, while a tiny shrew's heart goes at the amazing rate of over 600 beats every minute. The elephant will step to one side with slow deliberation compared with a small sparrow on the window ledge with its lightning movements. We can combine the concept of the time required for a life cycle and the time required for rapid physiological processes in an interesting way. A shrew will live only a year or two, but an elephant will average 40 to 50 years; yet they have one thing in common: the total number of heartbeats they have in their whole lifetime will be approximately the same. So life for the small beast goes faster because its engine is racing along compared to the larger beast, and the total budgets for their actions are the same.

Evolutionary time is another time scale in the realm of life. Now we are no longer dealing with one generation, but with a great series of generations going way back in time to the beginning of life on earth. We are no longer dealing with minutes or a few scores of years, but with millions, and even billions of years. We find in the fossil record an era when all life consisted of single cells, which later were followed by simple multicellular organisms, leading ultimately to all the great variety of animals and plants that we see on the earth today. This has been an exceedingly slow and an exceedingly grand evolution that has taken up a vast quantity of time—so much so that it is difficult to comprehend the magnitude of geological time. And we know that even these time spans are modest compared with those of the astronomer, who thinks in terms of light-years.

So what does a biologist think of this second millennium? It is too short a time for major changes in evolution, but time enough for many generations. Every 1000 years will allow some 50 human generations, but the shrew will have a new generation each year, which means a 1000 each millennium. So for slime molds and shrews the second millennium has meant waiting impatiently for a huge number of generations, while for elephants and ourselves—the wait barely tries our patience.

The important thing to note here is that size differences affect not only physiological processes, such as the speed of contraction of muscles, but these very effects are another facet of longevity, that is, the average time an animal lives. This is shown by the rather startling statement that the total number of heartbeats in the life of a hummingbird will be very roughly the same as that of an elephant or a whale. Increase in size means all life processes go more slowly and therefore take longer to happen. And this is all the result of living things clinging to efficiency rather than clinging to preserve the volume/surface ratio no matter what the size. Efficiency is the prime result of natural selection and by slowing down the metabolism with increase size, every aspect of the organism's existence will be affected by time, from the speed its motor is running to how long it will run. Time and size are inseparable for the great diversity of the animals and plants on this earth.

Note that all the processes we are considering here involve time. So if one divides the life span by the number of heart beats of an animal, it is time over time: another dimensionless number. And like the B/D ratio it remains roughly constant regardless the size of the animal. This will also be true of metabolic rate.

SIZE AND METABOLISM

One of the most discussed of all aspects of the size of organisms—and here we mean primarily animals—is the way size affects the rate at which the internal motor runs, that is, the metabolism.[29] Large animals obviously need more total energy simply because they are bigger; after all, a huge truck uses more gasoline than a minicar driving over a given distance. While this is undeniably the case, it is also true that there is an important difference between large and small animals. The rate at which they burn their fuel—be it sugar or some other food—in an ounce of flesh differs greatly: an elephant consumes much less energy-providing food in that ounce than does a mouse.

Why should that be so? The simple answer is that the components needed to keep the motor running have to enter by diffusion: the sugars, the amino acids, and all the other elements of energy-providing food must come in through the surface of the gut. And oxygen is needed to combust that food in the cells of the body and turn it into useful energy such as muscle contraction; it also must come in through a surface, that of a lung, or in the case of fish, a gill. So the running of the living motor is constrained by the extent of the organism's surfaces, by the square of the linear dimensions. On the other hand, the controlled combustion that takes place within the cells is limited by the number of cells making up an animal, that is, its total weight, which varies as the cube of the linear dimensions.

As we have seen from the earlier discussion of the effect of size on the shape of organisms, size is expressed as weight or volume, and when it increases in comparing two animals of different sizes, then surface structures increase at an even greater rate to keep up. However, there is a paradox: if the surface increase did keep pace exactly with volume, so that the ratio of surface to volume remained constant, then one would expect the metabolism for each cell to remain constant, regardless of the size of the animal it inhabits. Clearly this is not the case: each cell of an elephant combusts its fuel, its food, at a slower rate than that of a mouse.

Conventionally the relation between metabolism and size is expressed allometrically. That is,

$$y \text{ (metabolic rate)} \propto x \text{ (body weight)}^{a \text{ (constant)}}.$$

As before, this relation can be expressed logarithmically:

$$\log y \propto a \log x.$$

If this information is now plotted on a graph, it will give a straight line whose slope will equal a (fig. 31). This curve, the so-called mouse-to-elephant curve, was the brainchild of Max Kleiber in the 1930s, and he calculated that the slope was about 3/4 or 0.75. This tells us that in larger animals the extent of the increase of the surface does not keep pace with the increase in volume.

Following Kleiber there have been many others who have confirmed this mouse-to-elephant curve and extended the observations to include cold-blooded vertebrates,

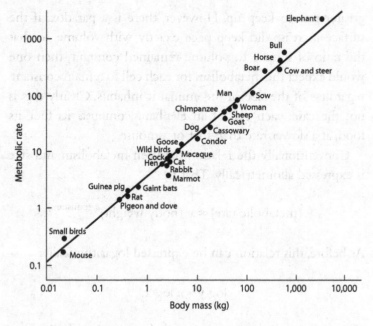

Figure 31. The "mouse-to-elephant curve," showing the relation of metabolic rate to body weight. (From Schmidt-Nielsen, *Scaling: Why Is Animal Size So Important?*, 1984, after Benedict, 1938)

invertebrates, and even down to small bacteria. There is a big literature on the significance of the allometric exponent of 0.75 and the variations of this slope and the theory that must lie behind it. The arguments have been ingenious, but it is a topic that is beside the main point I want to make here. I am looking simply for the explanation of why the slope does not equal 1. Why and how does an animal manage to keeps its motor going if its surfaces that absorb oxygen and food lag with size increase?

The why, like so many things in biology, can be answered in terms of Darwinian natural selection. We have gone over some of the arguments why there has been a selection for size increase, but there is also a constant selection, no matter what the size, for greater efficiency. Any animal must function in a way so that it is successful in producing offspring, and that may not just involve its size, but all aspects of it functioning: its physiological efficiency.

This brings us to the "how." To begin with, it is reasonable to assume that if a way can be found to use less fuel, less food and oxygen, and still function effectively, this will be an advantage. If an elephant's cells were burning its food with the same intensity as those of a mouse, it would require vastly more food than an elephant consumes. Furthermore, combusting food generates heat, and in order to keep its temperature at a reasonable level it would need a far greater surface for dissipating the heat, for cooling. Of course, if the surface/volume ratio is kept at 1, it would indeed have a greater surface, but it would no longer look like an elephant but have great convolutions of its skin and internal organs, rather like some sort of monstrous walnut. The solution is to slow down its metabolism just enough so that it can exist and look like an elephant. This is the main way an animal can and does manage size increase and remain efficient. That is perhaps not a sufficiently strong way of putting it: a larger animal could not even exist unless its cells had a reduced rate of metabolism. It would either starve or burst into flames, or both.

One consequence of the differences of metabolism for different size organisms is that small warm-blooded animals must eat constantly to keep their intense internal fires burning. In the case of a shrew, if it cannot find food there will be irreversible internal damage after a few hours and it will die of starvation, while we humans can go on a hunger strike and survive many days without perishing. Presumably elephants can go for even longer, although the information on how long is a bit less reliable because they are wise enough not to go on voluntary hunger strikes. There are other metabolism-related consequences, such as longevity, that we will examine shortly.

Let us return for a moment to *Gulliver's Travels* and consider where the various-size characters might fit on the mouse-to-elephant curve (fig. 31). Lilliputians are roughly the size of a mouse, Gulliver is already on the curve, and the Brobdingnagians are roughly the size of an elephant. This means that, like a mouse or shrew, the Lilliputians' heart is racing, while by comparison the heart of a Brobdingnagian pumps at an exceedingly slow rate. Furthermore, the movements of a Lilliputian will be as rapid as a blink, while those of a Brobdingnagian will be deliberate and slow. The differences in their sizes means big differences in all their activities.

SIZE AND SONG

One of the most curious effects of size is on the noises animals make, from speech to songs, from alarm cries to love calls; every form of animal communication. Sound does, after all, involve time for it depends on frequency. The larger the animal, the lower the frequency, the lower the sound. In the case of human beings, large individuals tend to have deeper voices; we expect a basso profundo to be a bigger man than a tenor, although this often turns out not to be true. It has to do with the length of the vocal chords: the longer they are, the deeper the voice. Think of the range of sounds of a violin compared with that of a huge bass viol whose longer strings can make much lower, very profundo sounds.

The point is well illustrated in frogs, where the larger males have a deeper croak that is particularly alluring to the females. There was a flurry of excitement at one time because it was suggested that if a small male sat in very cold water it would make his voice deeper for it would affect the muscles in his sound box: the undersized were freezing to death, all for love. Whether or not this romantic idea is true is uncertain, but the thought is rather nice.

The sound range of different animals has only been appreciated in the last sixty years. We are intimately aware of what human beings can hear, all the way from the piccolo to the boom of a kettle drum in a symphony orchestra, and for centuries we were totally unaware that we could not hear the sounds of some animals because they were too high or low pitched for our human ears. The first surprise came when it was found that bats made a whole spectrum of sounds above the upper limits of what we could hear. Not only that, but they used those sounds to "see" their environment: they sent out the sound signals and then they caught them as they bounced back to their ears from surfaces or objects. This was called *echolocation,* and with this amazingly sensitive system bats could catch tiny insects in the dead of night.

There has been an interesting recent report on the differences between large and small forms within one species of bat, the long-eared horseshoe bat of Southeast Asia and Australia.[30] As one would expect, the larger ones produce lower frequency sounds than the smaller ones with the result that they eat different size foods: the smaller bats can capture the smaller insects that the larger ones cannot find with their echolocation system. Because of these differences, the size forms may have become, or are becoming, separate species entirely because of their size difference and its effect on their vocal cords and the pitch of their calls.

This phenomenon was first revealed using instruments that could hear and record the high sounds to which we are deaf. This is another example where, like the microscope and the telescope, we can extend our faculties to not only see beyond our visual capacities, but our hearing capacities as well. Those sound recording devices also helped us at the other end of the scale to hear sounds at a very low register that we did not know existed. It turned out that whales made sounds of very low frequency, and they, like birds, used those sounds for courtship, for signaling between parents and young, and for disclosing their location. These sounds are in our hearing range, and some of the whale songs are quite haunting; at one time, whale music records were very popular. One trick I found very interesting was to play the whale song recordings much faster than normal; doing so made the low-frequency sounds higher. Now they no longer sounded like whales, but like the song of small, chirpy birds, making the point that size is indeed correlated with the frequency of sound waves.

A parallel story is now known for elephants. In the 1970s the zoologist Katy Payne was gazing at some elephants in a zoo and noticed that periodically the folds around their mouth would move as though they were exhaling, and she wondered if it was a sound she could not hear.[31] With the help of a microphone and a recording instrument she found that indeed they were making very low-frequency sounds inaudible to us. She then transferred her studies to Africa and eavesdropped on quite remarkable conversations of elephants in the wild. One of the things she noted, and indeed this is equally true for whale calls, is that low-frequency sounds, unlike high-frequency ones, will travel great distances. So elephants—or whales—that are miles apart can keep in touch.

The lesson from all this is that each of us hears in a frequency range corresponding to our size: bats are small and elephants and whales huge and we are somewhere in between. The same must be true for Gulliver: the Lilliputians must have squeaked like a mouse and there is a good chance that neither they nor Gulliver himself could even hear the speech of the Brobdingnagians.

SIZE AND GENERATION TIME

One of the consequences of an increase in size is that it takes longer to build a big animal or plant than it does a smaller one. Let us examine this fairly obvious point in more detail. For living things, the building process means growth, so the period of size increase is the period of development from the single-cell stage to that of maturity when it in turn can start a new generation. This period is the part of the organism's life cycle called the *generation time*. In human beings the generation time is about twenty years so that in a century one would expect about five generations for *Homo sapiens*. On the average it takes us that much time to grow up from conception to adults that produce children. We can reproduce at an earlier age, as all those teen-age pregnancies attest, nevertheless twenty years is closer to the norm. The timing of maturity in human beings is not a perfect match with the ability to produce egg and sperm. It is true our growth stops when we are about twenty and the growth zones in our long bones congeal, so maturity occurs at roughly the same time we begin, on the average, to have children.

There is a great difference between our generation time and that of smaller and larger organisms. At one extreme, bacteria—which are single cells—may divide every half hour. Compared to us, the number of their generations that can be packed into one of ours is positively enormous. On the other hand, a giant sequoia needs sixty years for a generation—it takes that long for it to produce cones with seeds. We can produce three generations during their one. That there is a clear relationship between the size of an organism and its generation time can be shown on a graph that plots the length of a plant or animal against its generation time (fig. 32). Admittedly there is considerable variation between species, but in a general sort of way, if one doubles the length of an organism, the time it takes to produce a generation doubles as well. No doubt there are many reasons why all species do not fall exactly on the line. It might have to do with the particular ecological conditions to which they are adapted; it may have to do with the way they are constructed; or, in the case of higher animals, it may even have to do with differences in behavior.

There is nothing that is not plain in this relationship. The bigger the plant or animal, the more time is needed to build it; there is no way of getting around this very obvious fact.

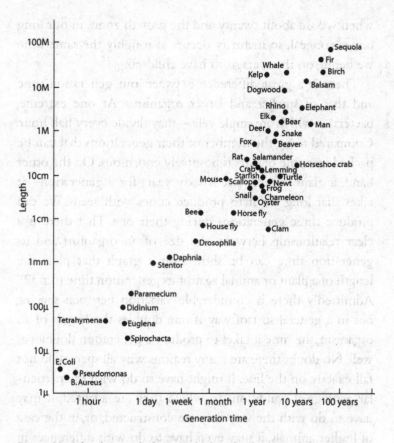

Figure 32. A log-log graph showing the length of an organisms at the time of reproduction plotted against its generation time. (From Bonner, *Size and Cycle*, 1965)

It applies to the nonliving world as well. A rowboat might be built in a day, but building an aircraft carrier will be a long-term enterprise. One need only think of the immense number of parts that need to be put into place in a carrier compared to the few screws required to construct a rowboat; no wonder there is a big difference in their construction time. Similarly, if one compares the time it takes for a minute alga made up of a few hundred cells in all to grow to its adult size, with that of a comparatively huge human being that has more than a trillion cells, it is not surprising that the alga can come to reproductive maturity in a matter of days, while we take a number of years before we can manufacture egg and sperm. Size increase or growth in living organisms consists largely of cell divisions, and although most cells in multicellular organisms do not divide as rapidly as bacteria, they may take hours, or sometimes days, between one division and the next. Since we both start as single cells it obviously takes a shorter time for the hundreds of cell divisions of the alga to mature compared to our trillions; no wonder we take longer to grow up.

LONGEVITY AND SIZE

There is another important aspect of life cycles that correlates with size: larger animals and plants live longer. This means that over and above the period of growth, the organism may live for a long time after its first reproduction—after the end of its generation time. Some organisms, such as ourselves, provide good examples. After the cessation of growth at age 20 we might continue to live many more years, perhaps three- or fourscore years beyond the first.

Many organisms do not stop growing after they start reproducing. Trees, for instance, keep growing for many years after the first time they bear fruit. Fish and many reptiles do the same. Being a fisherman, I cherish the continued growth of fish; I might someday still catch the ancient giant I've dreamed about. One summer some friends who rented a boat to sail among the Hebrides in Scotland came to a small fishing harbor. Looking over the boats was a statue of Jesus with his arms spread out, blessing the fleet. A fisherman saw them looking at it and said, "We call it, 'The one that got away.'" Even some mammals, such as the elephant, continue to grow for many years after they are capable of reproduction. In all such cases of continued growth, however, the rate of the growth slowly declines with age. Whether there is growth during maturity or not, the life span correlates well with size: bigger organisms live longer.

Life span is controlled in two basic ways. One, which is found in animals and in many other organisms including unicellular ones, is genetic: there are genes that control the life span. This is an intrinsic control mechanism; the other is extrinsic and is particularly relevant to the main message of this book. It is what controls the size of trees—it is what makes an old tree come to the end of its growing. There is good evidence that a tree starts declining in its growth when it becomes so tall that the water and nutrients can no longer effectively reach the growing tips of the outer stems. This has been shown in a recent study in which four species of trees were measured. For each species, the growth of the terminal stems of large trees in the field were compared with growth in cuttings from those same trees, and only the outer stems on the large trees showed a decline. If young shoots are grafted into the crown of an old tree, they will decline in their growth rate in their new environment comparable to that of their neighboring stem tips.[32] Parenthetically, this probably explains why Bonsai trees, which are continually trimmed to keep them small, are known to live for exceedingly long times—one tree is reputed to be over seven hundred years old. Tree growth is a perfect example of how size can exert supreme control over life.

If one plots the size of an organism, in this case mammals, against their life span (in the sense of the maximum length of life known for individuals of a species) one will have a curve that will reflect what I have just described, that larger organisms live longer (fig. 33a).[33] Note that the scatter of points is considerable, much as they were with generation time and size in figure 32, although in both cases the trend is clear. As before, there are no doubt many reasons for the scatter, but none of them overshadow the obvious trend. The puzzling fact is that for the same mammals shown in the figure,

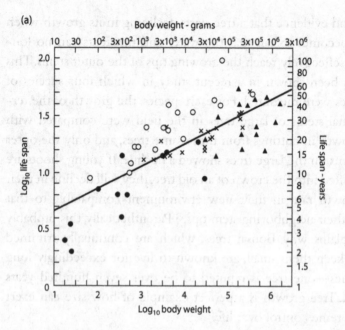

Figure 33. (a, above) A log–log graph showing the relation of life span to body weight for mammals. Open circles are primates and lemurs; solid circles are rodents and insectivores; crosses are carnivores; solid triangles are ungulates and elephants; the open circle at the top is a human being. (b, facing page) A graph for the same animals, in which life span is plotted against brain weight. (From G. A. Sacher, *CIBA Foundation Colloquia on Aging* 5 [1959]: 115–141)

if instead of body size the brain size is plotted against life span, the scatter is greatly reduced (fig. 33b). Why can this possibly be? Why is brain size a more reliable indicator of life span than body size? Perhaps it is because body size can be greatly

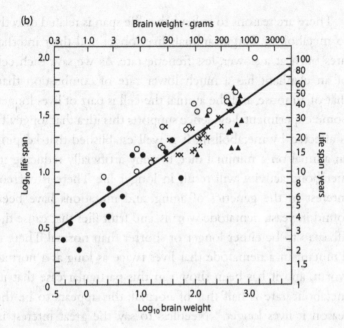

affected by environmental conditions, such as the food supply, while the brain is permanently encased in a bone prison.

One point is above the curve (the top circles in figs. 33, a and b) and is far above the average. This is the point for human beings—we have an unusually long life span for our size. It has been suggested that the reason for this is that our comparatively large brain means that our elders may help the general welfare of the group through their storage of wisdom, which has resulted in selection for a long postreproductive span of years.

There are reasons to believe that life span is related directly to metabolism. Larger animals live longer and their internal fires burn at a slower, less frenetic rate. As we saw, each cell of an elephant has a much lower rate of combustion than that of a mouse, and the animal the cell is part of lives longer. Some experimental evidence supports this idea that longevity is affected by metabolism. It is well established that keeping an animal on a minimal diet, thereby artificially reducing its metabolic activity, will result in longer life. There is current interest in the genetics of aging, and mutations have been found in yeast, nematode worms, and fruit flies that cause the life span to be either longer or shorter than normal. There is a mutant in a nematode that lives twice as long as a normal worm, and it has been shown in this particular case that its metabolic rate is half that of normal; this appears to be the reason it lives longer.[34] Needless to say, the great interest in longevity comes from our permanent fascination with the idea of extending human life and our endless quest for the fountain of youth. It is doubtful that one could extend human life by altering our genes, and even if one could it might be that we would end up with people who were extremely slow and lethargic, which sounds like too high a price to pay for becoming a Methuselah.

It is also instructive to think of the mouse-size Lilliputians. They would live only a year or two compared to Gulliver's sixty to a hundred. If we compare the huge Brobdingnagians with Gulliver, we would find that these giants—larger than elephants—might live about twice as long as we do.

SIZE AND SPEED

As we have seen, this reduction of metabolism with size increase is intimately tied in with the fact that a larger animal does everything at a slow pace. A hummingbird's wings move so fast that one cannot see them, while a great blue heron gracefully waves its wings in what, by comparison, appears to be slow motion. We can observe this relation between size and movement in our own bodies. The very small muscles in our eyelids that are responsible for blinking can contract in a split second, while by comparison we move our legs slowly. In this case the difference is the direct result of the size of the muscle.

In 1950 an important paper was published by the noted English physiologist A. V. Hill in which he examined this very point.[35] He concluded that animals of exactly the same proportions should run at the same speed: for example, a mouse and some similarly constructed large mammal ten times taller would tie in a race. The mouse's legs would make a step that is one-tenth of the length of a stride of the larger animal, but it would move ten times faster, so they should run at the same speed. As examples he compared a whippet, a greyhound, and a horse whose maximum speeds were 34, 38, and 42 mph, respectively.

Hill's general principle is particularly apt for Jonathan Swift's people of three different sizes. Each is ten times taller than the other: there are three orders of magnitude from Lilliputians to Brobdingnagians. Moreover, they are similarly constructed; all are shaped and proportioned like human beings, however improbable that might be, as we have seen. This means they perfectly fulfill Hill's conditions and therefore they should be able to run at exactly the same speed. One step of a Brobdingnagian would take the same amount of time as ten rapid steps of Gulliver, and one hundred lightning steps for a Lilliputian. The latter's small muscles would vibrate like the wings of a hummingbird, while each step of the giant would be stately and slow, but they would all reach the finish line together. The principle is undoubtedly sound, but clearly it is not the whole story; larger animals do run faster, as one can even see from his example of racing speeds in dogs and horses.

Some years ago in looking for the possible advantages an animal might have by being larger I searched the literature for information on the speed of various animals, not only those that ran, but also those that swam or flew. It turned out to be an interesting exercise in which the conclusion is that, in general, bigger meant faster regardless of whether the animals were similarly constructed or not (fig. 34).

The smallest runners I could think of were mites, but I could not find anything on their running speed. I had met Paul Winston at the University of Colorado who works with

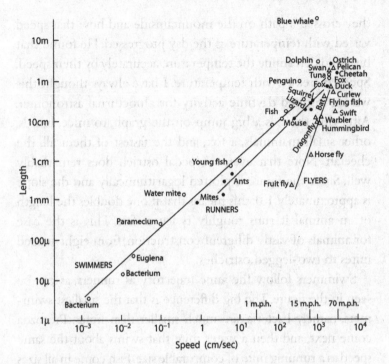

Figure 34. The speed of various animals—swimmer, runners, and flyers—plotted against their length on a log-log graph. (From Bonner, *Size and Cycle*, 1965)

mites, so I wrote him to ask if he had any information. He said he did not, but that he and his colleagues would run some mite races of different-size mites, and the results are shown on the figure. The next bigger runners were ants measured by the distinguished astronomer Harlow Shapley on Mount Palomar.[36] He was interested in the speed of two species as they crossed a path on the mountainside and how that speed varied with temperature as the day progressed. He found that he could determine the temperature accurately by their speed. Speed increases with temperature. I have always thought this was an inspired daytime activity for a nocturnal astronomer. After this, there is a big jump on the graph, to mice, lizards, other small mammals, a fox, and the fastest of them all, the cheetah. Note that even a bipedal ostrich does remarkably well. Since the figure is plotted logarithmically and the slope is approximately 1.0, this means that if one doubles the length of an animal it runs roughly twice as fast. This is the case for animals of vastly different construction, from eight-legged mites to two-legged ostriches.

Swimmers follow the same trajectory as runners, as can be seen in the figure. The big difference is that the smallest swimmers, namely bacteria, are much smaller than mites. Protozoa come next, and then a water mite that swims about the same speed as a running mite of comparable size. Fish come in all sizes from larvae up to mature tuna and the related wahoo. All seem pretty close to the line, with the possible exceptions of the penguin and dolphin; the only animal that clearly does not fit is the blue whale; the reason for this divergence is poorly understood.

The smallest flyer about which I could find information on speed was the fruit fly (*Drosophila*), so important as a laboratory animal. There are much smaller flying insects, such as fairy flies (which are parasitic wasps; see figs. 6 and 12), but no one seems to have clocked their speed. Going up the curve in the figure, we see that horse flies are faster than fruit flies. Next we find hummingbirds and dragonflies, a bat, a warbler. A swift matches its name, and it is fast for its size, while a flying fish, a curlew (whimbrel), and a duck fall on the line. The fastest flyers are large: the swan and the pelican.

If we stand back and look at some of the implications of this graph, for starters, it is obvious that, like running speed, swimming speed is proportional to length; if the length is doubled, so is the speed. The advantage of increased size for a flying animal is less dramatic, although it is clearly there. The problem encountered in flying has to do with the extra horsepower needed for flight compared to swimmers and runners.

Another particularly interesting fact that is revealed by the figure is that the three lines join in roughly the same area. This means that the fastest animals are in the general neighborhood of one meter in length, and this is true for swimmers, runners, and flyers. A great excess in length, such as found in the blue whale, does not produce any further advantage in speed.

SIZE AND SPEED FROM THE COSMOS TO ELEMENTARY PARTICLES

Motion is everywhere and not just in living organisms. Planets and stars race about at great speeds, as do atoms and elementary particles within those atoms. Everything in the universe moves.

If their size and their speeds are compared to those of living organisms, the latter fall into an intermediate range between the largest and smallest physical entities (fig. 35). Why electrons and other subatomic particles rush around at such incredible

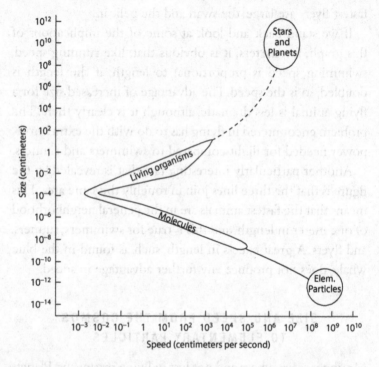

Figure 35. A diagrammatic graph on a logarithmic scale showing the relation of speed of movement to size, from elementary particles to celestial bodies. (From Bonner, *The Evolution of Culture in Animals*, 1980)

speeds, as indeed do celestial bodies, is a great mystery to me. I am neither a physicist nor an astronomer and therefore it is fortunate that the matter is beyond the province of this book. However, let us examine the left-hand part of the figure more closely: the world of molecules and living organisms.

We know from those early science classes at school that molecules move and that their motion is affected by temperature. If they are in a solid state, as are water molecules in ice, they are closely packed and their motion is greatly restricted. At warmer temperatures, when the ice turns into liquid, the molecules are released from prison and now move around with agility; at still higher temperatures they turn into water vapor, a gas, and now those molecules zip around at a furious pace with great distances between them. In fact, the three states—solid, liquid, and gas—are defined by the movement of molecules, and those states are determined by the temperature. It is this same molecular motion that is responsible for diffusion, as we discussed earlier: diffusion of oxygen through the lung surface or the diffusion of amino acids or sugars through the gut surface of animals. Those processes are totally dependent upon molecular motion.

One feature of this motion is that the smaller the molecule, the faster it moves, as is clearly shown in figure 35. Note on the graph that the largest molecules and the smallest bacteria meet: the point of slowest motion of all of life and of all non-living physical bodies.

In this region of minimal speed we are in the realm of Brownian motion. In 1827 Robert Brown, an English botanist, was looking inside some plant pollen and saw small bodies bouncing about. He correctly pointed out that the movement could not be a manifestation of life because the pollen was at least one hundred years old and therefore totally dead. With the rise of physical chemistry that followed as the century progressed, it became evident that molecules in the solution were bombarding the small visible particles in the liquid. Those molecules were too small to be seen in a light microscope, but their effect could be observed in the movement of larger visible bodies that were being rammed into, giving them a drunken, zigzag motion. The subject culminated in a paper by Albert Einstein published in 1905 in which he established the theoretical basis of the phenomenon.

In terms of size and speed, large molecules are more or less in the same league with living bacteria, but from an evolutionary point of view the gap is enormous. The long, slow origin of life must have spanned many millions of years. However, here we are only concerned with speed and size, and large molecules and small bacteria move at roughly the same pace, a pace that is the slowest of everything that moves in the universe. Bacteria move, as was discussed earlier, by an elaborate and clever rotary propelling of flagella; molecules, on the other hand, do not have a motor but are propelled by inner energies that are part of their fundamental makeup. The larger the molecule becomes, the slower it moves. Motile organisms are just the opposite, for as we have seen, the bigger they are, the faster they travel. (The living organism balloon drawn on the figure encompasses all of the previous figure [fig. 34], the one that shows the speeds of different-size animals.)

What came first: did the invention of a means of propelling the organism come first and cause the evolution of size increase, or the reverse? Selection for size changes had to come first and become the prime mover, for if the selection for movement followed size increase, then one would expect different mechanical ways of dealing with a particular size, and that is exactly what has happened. The smallest motile beasts, namely bacteria, have the extraordinary mini-rotary motor that can effectively move in the world of extremely low Reynolds numbers; in larger single-cell organisms, flagella do not rotate but contain contractile proteins that allow them to become active whips; in larger unicells, cilia beat like oars to move the cell (fig. 11); this is followed by the invention of muscle and bones to deal with running, swimming, and flying in larger animals. The progression of size during the course of evolution leads to a series of different inventions for locomotion, each appropriate for the dimensions of the organism.

Obviously, this effect of size only applies to organisms that move, and there are many that do not: many species of bacteria, algae, fungi, many invertebrates such as sponges and corals, and all vascular plants. Only some kinds of organisms need to move. In some of the larger immobile forms, the gametes may be motile, single-flagellated cells. Even the largest animals, elephants and whales, have motile sperm, but note that the sperm's mechanical means of locomotion is appropriate for their small size: the single-cell sperm of all larger animals do not move using muscle and bones, but they have a flagellum. Here, too, size dictates the locomotion machinery, and the reason that this is so is purely physical; a whale could not manage to move at all if its bulk had to rely on a furry coat of beating cilia. Again we see evidence of the overreaching role of size in all of life.

CHAPTER 7

ENVOI

Size and the actual organisms that are measured are rather like shadow and substance. The organisms are material objects. In correlation with their size, they come in many shapes and have different physiologies, degrees of internal complexity, generation times, life spans, speeds, and frequencies to their songs. A beast's size is like a shadow: it has no substance. It is simply a statement of how much matter makes up a particular living entity.

What is remarkable is that although size is no more than a shadow, no more than a description, a property, it nevertheless exerts enormous power over the living world; it drives the form and function of everything that lives. How can this be so?

The root reason is natural selection. The size of an organism is under constant selective surveillance. Nature is made up of a vast network of size niches, and all living forms are always faced with the possibility that it may be advantageous to have their descendants become larger or smaller. If they are animals, size may be important for avoiding predators, or catching prey; for plants it may be important for success in the competition to catch the rays of the sun for photosynthesis. In microorganisms the most obvious advantages in becoming larger may also involve greater speed to pursue prey or escape predators. Becoming larger by multicellularity may mean more effective dispersal of spores, or more effective digesting of food, where the presence of more digestive enzymes are insured by increasing numbers of cells. The list could go on, as I have argued in much of this book. The key point is that size changes—increases or decreases—will have advantages or disadvantages and will be encouraged or discouraged by natural selection. Selection is the motive force for size change.

And if the size is changed there will be a myriad of consequences, especially if it is increased. It will mean various constraints bear down: size is volume, yet life's activities require the appropriate surface to go with the volume and the result will be different shapes for different sizes. This will have many ramifications affecting properties such as metabolism and locomotion. It is particularly interesting that as the volume increases, there is an increase in the division of labor in the form of the number of cell and tissue types. Unless these and many of the associated changes occurred, larger organisms simply could not manage. They would not exist. Size increase has forced all these changes. The shadow is not really just a shadow: it is much more. Even though it is not substance, it is a dictator that holds complete sway over what on organism will look like and how it will function. Size rules life.

In the cases of size decreases, size plays a somewhat different role—it is no longer the absolute dictator. This is because having less bulk may permit a decrease in the structures that are involved in matters of strength or surfaces for diffusion, but they do not necessarily require it. It could be that with continued size decrease, they may not have room for all the structures their larger ancestor possessed and may ultimately have to shed some of them; clearly size decrease is far less demanding than size increase.

If we stand back even farther, we see a principle here that goes beyond life. In the inanimate physical world, size has the same effect as it does on living organisms: Galileo made this clear right from the start. A small pendulum swings faster than a large one; a large boat such as an eight-oared shell will move faster than a one-man single. I pointed out earlier that it takes longer to build an aircraft carrier than a rowboat, but also that the larger ship has a vastly larger division of labor in its parts. The proportions of bridges or of buildings must differ with size increase because of the weight-strength relationship. The engineer who builds airplanes, engines, or bridges must make size one of his prime concerns; if it is neglected, all his endeavors will fail.

One could say that what makes life unique is natural selection that culls so that only the efficient constructions survive, while the equivalent for human constructions is whether or not they work. Mechanical or physical efficiency in engineering structures simply has a different method of selection, just as the selection forces of economic efficiency governed the division of labor in human societies. Size is ubiquitous and casts its shadow on all material things, both the living and the nonliving.

Putting all these matters concerning size together leads to another insight. The size rules that have been laid out show that there is a connection between strength, surface activities, division of labor, and all activities that involve rates of processes such as metabolism, generation times, longevity, speed of locomotion, and even the abundance of organisms in nature. What connects them all is size. A change in weight will either require, or be correlated with, a change in strength, in surfaces, in the division of labor, in all the time-related rate processes, and even in the density of the distribution of organisms in nature. Each of the five size-related categories is interdependent with all the others: one changes and they all change. It is much the same thing as one finds in the gas law where changes in temperature will change the pressure or the volume of a gas: a similar kind of connectivity is found in the size rules.

All the thoughts in this book reflect the human preoccupation with size. This is so for Galileo and all the scientists who have followed him up to this very minute as I write these words. It is equally so in literature from Sinbad the Sailor through the delights of *Gulliver's Travels*, right up to many artists and writers today. And one can add to them a multitude of aspects of human existence: architecture, engineering, business, transportation, and innumerable others. There is no corner of human endeavor and human thought that escapes the tentacles of size. Couple size with the evolution of living organisms—and this becomes the book I have written.

NOTES

1. Frye, T. C., G. B. Rigg, and W. C. Randall. The size of kelps on the Pacific coast of North America. *Botanical Gazette* 60: 473–482 (1915).

2. Ferguson, B. A., T. A. Dreisbach, C. G. Parks, G. M. Filip, and C. L. Schmitt. Coarse-scale population structure of pathogenic *Armillaria* species in a mixed-conifer forest in the Blue Mountains of northeast Oregon. *Canadian Journal of Forest Research* 33: 612–623 (2003).

3. See Random Samples. *Science* 305: 472 (2004).

4. Galileo, Galilei. *Dialogues concerning Two New Sciences.* New York: Dover Publications, 1914.

5. Went, F. W. The size of man. *American Scientist* 56: 400–418. (1968).

6. For a more rigorous discussion of Reynolds number, see McMahon, T. A. and J. T. Bonner. *On Size and Life.* New York: Scientific American Library, 1983.

7. Purcell, E. M. Life at low Reynolds number. *American Journal of Physics* 45: 3–11. (1977).

8. The pioneer of allometry was Julian Huxley with his path-breaking book, *Problems in Relative Growth.* London: Methuen, 1932.

9. Rensch, B. Increase in learning with increase in brain size. *American Naturalist* 90: 81–95 (1956).

10. Weir, A. A., J. Chappell, and A. Kacelnik. Shaping of hooks in New Caledonian crows. *Science* 297: 981 (2002).

11. de Waal, F. *Chimpanzee Politics.* New York: Harper and Row, 1982.

12. Byrne, R. W., and A. Whiten, eds. *Machiavellian Intelligence.* New York : Oxford University Press, 1988; II, New York: Cambridge University Press, 1997.

13. Fankhauser, G. The effect of changes in chromosome numbers on amphibian development. *Quarterly Review of Biology* 20: 20–78 (1945).

14. Fankhauser, G., J. A. Vernon, W. H. Frank, and W. V. Slack. Effect of size and the number of brain cells on learning of larvae of the salamander *Triturus viridescens. Science* 122: 692–693 (1955).

15. Borass, M. E., D. B. Seale, and J. E. Boxhorn. Phagotrophy by a flagellate selects for colony prey: a possible origin of multicellularity. *Evolutionary Ecology* 12: 153–164 (1998).

16. Limoges, C. Milne-Edwards, Darwin, Durkheim, and the division of labor: a case study in the reciprocal conceptual exchanges between the social and the natural sciences. In *The Natural and the Social Sciences*, ed. I. B. Cohen. Cambridge, Mass.: MIT Press, 1940.

17. Bell, G., and A. O. Mooers. Size and complexity among multicellular organisms. *Biological Journal of the Linnaean Society* 60: 345–363 (1997).

18. Kirk, D. *Volvox.* Cambridge, U. K.: Cambridge University Press, 1998.

19. Schaap, P., T. Winckler, M. Nelson, E. Alvarez-Curto, B. Elgie, H. Hagiwara, J. Cavender, A. Milano-Curto, D. E. Rozen, T. Dingermann, R. Mutzel, and S. L. Baldauf. Molecular phylogeny and evolution of morphology in the social amoebas (submitted).

20. Bonner, J. T., and M. R. Dodd. Aggregation territories in the cellular slime molds. *Biological Bulletin* 122: 13–24 (1962).

21. Waters, C. M., and B. Bassler. Quorum sensing: cell-to-cell communication in bacteria. *Annual Review of Cell and Developmental Biology* 21: 319–346 (2005).

22. Donner, J. *Ordnung Bdelloidea*. Berlin: Akademie-Verlag, 1965.

23. Valentine, J. W., A. G. Collins, and C. P. Meyer. Morphological complexity increase in metazoans. *Paleobiology* 20: 131–142 (1994).

24. McCarthy, M. C., and B. J. Enquist. Organismal size, metabolism and the evolution of complexity in metazoans. *Evolutionary Ecology Research* 7: 681–696 (2005).

25. Hölldobler, B., and E. O. Wilson. *The Ants*. Cambridge, Mass.: Harvard University Press, 1990.

26. Dawkins, R. *The Selfish Gene*. New York: Oxford University Press, 1976.

27. Peters, R. H. *The Ecological Implications of Body Size*. Cambridge, U. K.: Cambridge University Press, 1983.

28. *Tricycle* 34 (Winter): 64–65 (1999).

29. For an excellent treatment of metabolism and body size, see K. Schmidt-Nielson, *Scaling: Why Is Animal Size So Important?* Cambridge, U. K.: Cambridge University Press, 1984.

30. Kingston, T., and S. J. Rossiter. Harmonic-hopping in Wallacea's bats. *Nature* 429: 654–657 (2004).

31. Payne, K. B., M. Thompson, and L. Kramer. Elephant calling patterns as indicators of group size and composition: the basis for an acoustic monitoring system. *African Journal of Ecology* 41: 99–107 (2003).

32. Peñuelas, J. A big issue for trees. *Nature:* 437: 965–966 (2005); Mencuccini, M., J. Martinez-Vilalta, D. Vanderklein, H. A. Hamid, S. Lee, and B. Michiels. Size-mediated ageing reduces vigour in trees. *Ecology Letters* 8: 1183–1190 (2005).

33. Sacher, G. A. The relation of lifespan to brain weight and body weight in mammals. *CIBA Foundation Colloquia on Aging* 5: 115–141 (1959).

34. Van Voorhies, W. A., and S. Ward. Genetic and environmental conditions that increase longevity in Caenorhabditis elegans decrease metabolic rate. *Proceedings of the National Academy of Sciences* 96: 11399–11403 (1999).

35. Hill, A. V. The dimensions of mammals and their muscular dynamics. *Science Progress* 38: 209–230 (1950).

36. Shapley, H. Thermokinetics of *Liometupum apiculatum* Mayr. *Proceedings of the National Academy of Sciences* 6: 204–211 (1920); Note on the thermokinetics of Dolichoderine ants. *PNAS* 10: 436–439 (1924).

INDEX

acromegaly, 13
Acytostelium, 89–90
Aepyornis, 20
albatross, 18
Alice in Wonderland, 11
allometry, 50–56
amoeba, 26
Arabian Nights, 20
Aristotle, 1, 62
artificial selection, 99–100
Australopithicus, brain size, 55

B/D ratio, 114–115
bacteria, discovery of, 9; 27;
    swimming, 45, 48; generation
    time, 116, 129; speed of, 140
Baldauf, S., 89
Barnum, Phineas T., 12

Bassler, Bonnie, 93
bat, long-eared horseshoe, 126
Bonsai trees, 133
Brachiosaurus, 16, 73
brain size, 52–60
bridges, construction of, 1, 149
Brobdingnagians, 13–15, 33–34,
    39, 77, 124, 128, 138
Brownian motion, 144
Bursaria, 26, 69

Caledonian crows, 57
Carroll, Lewis, 1
cell size, 69–72
cell type number, as measure of
    complexity, 81–83
chimpanzee politics, 58
Chlamydomonas, 84–85

9 780691 254401